MILITARY BALLOONING.

Rediscovery Books

❖

Reproduced by kind permission of the
Royal Geographical Society

Published by
Rediscovery Books Ltd
Unit 10, Ridgewood Industrial Park,
Uckfield, East Sussex,
TN22 5QE England
Tel: +44 (0) 1825 749494
Fax: +44 (0) 1825 765701

This edition © Rediscovery Books Ltd 2006

To find out more about Rediscovery Books
and its range of titles visit
www.rediscoverybooks.com

Published in association with

Royal Geographical Society
with IBG

Advancing geography
and geographical learning

The **Royal Geographical Society with IBG** was founded in 1830 to advance geographical science. Today it supports geographical research, promotes geography in schools and through outdoor learning, in society and to policy makers. Geography connects us to the world's people, places and environments.

The **Rediscovery Books** series allow us to see how previous geographers and travellers understood and recorded the world.

In reprinting in facsimile from the original, any imperfections are inevitably reproduced and the quality may fall short of modern type and cartographic standards.

Printed and bound by CPI Antony Rowe, Eastbourne

PREFACE.

THE following Manual, written and compiled by Capt. B. R. WARD, Royal Engineers, will, it is hoped, prove useful not only as an *aide-mémoire* to Ballooning Officers, but also as a means of disseminating information throughout the Army generally regarding the use of Balloons.

Acknowledgments are due to Lieut.-Colonel C. M. WATSON, C.M.G., Royal Engineers, for the use of his valuable Ballooning Tables; to Major WILLOUGHBY VERNER, Rifle Brigade, for contributing the interesting chapter on "Free Run Reconnaissance"; to Capt. H. B. JONES, Royal Engineers, whose notes have been largely made use of throughout, more especially in the case of Chapter V, Section 2, "Instructions for the Conduct of a Free Run"; and to Lieut. A. H. W. GRUBB, Royal Engineers, for his chapter on "Photography from Balloons."

JAMES TEMPLER, *Lieut.-Colonel*
7th King's Royal Rifles,
Instructor in Ballooning.

SCHOOL OF BALLOONING,
    ALDERSHOT.
        1896.

MANUAL OF MILITARY BALLOONING.

Chapter I.
	PAGE
Introductory and Historical	5

Chapter II.
Manufacture and details of Ballooning plant	16

Chapter III.
Gas manufacture	28

Chapter IV.
Captive work..	33

Chapter V.
Free runs—
Section (1). Theoretical considerations	43
,, (2). Instructions for the conduct of a free run	50

Chapter VI.
On the employment of Balloons in War	59

Chapter VII.
Free Run Reconnaissance (by Major Verner, R.B.)	63

Chapter VIII.
Photography from Balloons	71

Appendix A.
Miscellaneous Tables	75

Appendix B.
Tables for facilitating Ballooning Calculations	104

CHAPTER I.

INTRODUCTORY AND HISTORICAL.*

The problem of aerial navigation has long been a fascinating one to scientists and enthusiasts alike, and very many ineffectual projects were put forward before the brothers Montgolfier solved the problem with their hot-air balloon in 1783.

One of the first of these projects was devised by an Italian Jesuit, Francis Lana, who, in a book published at Brescia, in 1670, proposes to construct four hollow copper spheres, each 25 feet in diameter, and $\frac{1}{225}$ inch thick, and he calculates (being ignorant of the pressure of the surrounding atmosphere) that if a vacuum were produced in them, they would rise from the ground with a total force of 1,120 lbs.

This idea of rising into the air by means of hollow copper spheres was, however, originally conceived some 400 years previously by our countryman, Roger Bacon, who states that the secret was known at the time to but one person besides himself. Lana's pamphlet inferentially points out the practical advantage that would arise from the power of thus moving through the air at will, for he earnestly prays God to avert the danger that would result from the successful practice of the art of aëronautics to the existence of civil government and of all human institutions. "No walls or fortifications," he says, "could then protect cities, which might be completely subdued or destroyed, without having the power to make any sort of resistance, by a mere handful of daring assailants, who shall rain down upon them fire and conflagration from the region of the clouds."

Another scheme of similar object was devised about 100 years later, by a Dominican friar, Joseph Galien, Professor of Philosophy and Theology at the Papal University of Avignon. His pamphlet, published in 1775 (not very long before the final invention of balloons), proposes to collect the fine rarefied air of the upper regions, far above the loftiest mountains, and to enclose it in a huge cubical bag constructed of stout sail cloth, the side of the cube being upwards of a mile in length. With such a vast machine it would be possible (he thought) to transport a whole army, and all their munitions of war, from place to place as desired.

In 1780 (only five years after Galien's project), the Montgolfiers made a variety of experiments, which resulted in the construction of their smoke balloons, thus laying the founda-

* The earlier portion of this historical chapter is taken from an interesting paper by the late Lieut.-Colonel Grover, R.E., published in Vol. XII of the *R.E. Professional Papers* for 1863.

tion stone, as it may be called, of the balloons of the present day. Their machine was composed of coarse linen, with a paper lining; it was pear-shaped, 75 feet high, and with a transverse diameter of 43 feet. The smoke of 50 lbs. of dry straw in small bundles, joined to that of 12 lbs. of wool, was found sufficient to fill it in 10 minutes, the white smoke produced from combustion having a specific gravity of 0·7 that of atmospheric air.

On the 5th June, 1783, M. Pilâtre de Rozier, the first aeronaut, and the Marquis d'Arlandes, made an ascent from Paris in a balloon, as described above, and after an adventurous run of 20 minutes, and a narrow escape of losing their lives, owing to the envelope catching fire, landed safely in the outskirts of the town.

As instances of the practical uses that the Montgolfiers proposed to derive from their invention, they suggested that "large balloons might be employed for victualling a besieged town, for raising wrecked vessels, perhaps even for making voyages, and certainly in particular cases for observations of different kinds, such as reconnoitring the position of an army, or the course of vessels at 25, or even 30, leagues distance."

Only a short time after the successful experiment of the Montgolfiers, viz., on the 1st December, 1793, a newly discovered gas (hydrogen) was used to inflate a balloon, by Professor Charles, a French physicist. Professor Charles thus laid the foundation of modern ballooning, and for a long time balloons inflated with hydrogen were known as Charlières, in opposition to Montgolfières, or hot-air balloons. Both terms are, however, now practically obsolete.

The French, by whom the actual idea of balloons was originally conceived and carried into effect, were also the first to discover the adaptability of their invention to practical purposes.

At the commencement of the revolutionary war, about 10 years after the production of the Montgolfier balloons, an Aerostatic Institute was formed by command of the French Directory (at the suggestion of Guyton de Morveau), in the École Polytechnique, and under its superintendence reconnoitring war balloons were constructed by a M. Conté, and supplied to each republican army in the field. The army of the Rhine and Moselle was provided with two, viz., the "Hercule" and "Intrépide;" another, named the "Céleste," was prepared for the use of the army of the Sambre and Meuse;. the "Entreprenant," for the army of the North; and a fifth was destined for the army of Italy.

That attached to the army of the Sambre and Meuse, under General Jourdan, was first used May 1794, by Captain Coutelle, at Maubeuge, before Mayence, in reconnoitring the enemy's works. This balloon, which was 27 feet in diameter, and took at first 50 hours to inflate, was retained to the earth by two ropes, and the aeronauts communicated their observations by throwing out weighted letters to the General beneath.

After this method of reconnoitring had been successfully practised four or five days, a 17-pr. gun was brought down to a neighbouring ravine, and (being thus masked) suddenly opened fire upon the balloon. Several shots were fired without effect, and the machine was then hauled down; but the next day the gun was forced to retire, and the reconnaissances were then carried on as before.

After two or three weeks, the balloon was moved to Charleroi, distant from Maubeuge about 36 miles. To save the expense and trouble of another inflation, it accompanied the troops at a sufficient height to allow the cavalry and baggage wagons to pass beneath, 10 men marching on either side of the road, and each man holding a separate rope attached to the balloon, which was thus retained at its proper elevation.

After making one observation on the way, the balloon arrived before Charleroi at sunset, and the Captain had time before close of day to reconnoitre the place with a General Officer.

Next day they made a second observation in the plain of Tumet, and, at the battle of Fleurus, which took place on the following day, June 17th, 1794, the balloon was employed for about eight hours, hovering in rear of the army at an altitude of 1,300 feet.

The Austrians after some time discovered it, and a battery was opened against the aeronauts, but they soon gained an elevation out of the range of the enemy's fire, and the information concerning the Austrians' movements (which they were enabled in this manner to supply to General Jourdan) contributed mainly, it is said, to the success of the day, the result of which was the loss to the Prince of Coburg and the allied armies of all Flanders, Brabant, &c.

This notable instance of the successful employment of a reconnoitring balloon is thus commented upon in the French history, *La Guerre de la Révolution de France*.—" Ce fut à cette bataille (Fleurus) que l'on fit, pour la première fois, l'essai d'un aërostat, avec le secours duquel le Général Jourdan put être parfaitement instruit des dispositions et des mouvemens de l'ennemi; ainsi, cette découverte, regardée jusqu'alors comme un objet de pure curiosité, dut être, dès cet instant, rangé parmi les inventions utiles."

Captain Espitallier, in his valuable book, *Les ballons et leur emploi à la guerre*, says, " Le ballon de Fleurus, l'Entreprenant, avait 27 pieds de diamètre, soit 8 mètres environ. Le jour de la bataille, il resta près de dix heures en l'air, et le Général Morlot, qui se trouvait dans la nacelle avec Coutelle, ne cessa pas un instant de tenir Jourdan au courant des moindres mouvements des Autrichiens."

The following is from a private journal kept by a Dr. Miers, of Hamburg, when on a visit to Paris:—" J'ai vu à Paris et à Meudon le Capitaine Coutelle, le même qui le 17 Juin, 1794, montoit le ballon qui dirigeoit la merveilleuse et importante reconnaissance de l'armée ennemie à la bataille de Fleurus, accompagné d'un Adjutant Général. Je lui ai parlé de son voyege aërien, pendant cette bataille, si décisive par suites, dont le succes est du en partie à cette expédition aërostatique d'apres le jugement unanime des personnes impartiales. Coutelle correspondit avec le Général Jourdan, Commandant de l'armée française, par les signaux de pavillon convenus."

The next battle that the French gained through the assistance of a balloon was near Liège, on the Ourte river.

As the Austrian officers afterwards said, "one would have supposed the French General's eyes were in our camp," for they were attacked at the critical moment of sending off their guns and baggage by the rear, the French (though occupying much lower ground than the Austrians) having been intimately acquainted with all their movements, by means of their balloon.

As a set off against the above testimonies it should, however, be stated that it would appear that both Generals Jourdan and Kléber reported unfavourably on the use of balloons, and that in consequence of this and of the failure of the balloon corps to achieve any results in Egypt, Napoleon disbanded the balloon establishments early in the present century. Captain Coutelle, however, puts a different complexion on the matter. In his report he informs us that the Aerostatic Institute was established in 1793, and abandoned on Bonaparte's return from Egypt in 1802. M. Conté, the director, had followed Bonaparte in this latter expedition, but the English having seized the vessel in which the apparatus for generating hydrogen had been embarked, the balloon was not employed in Egypt.

The French war balloons were inflated in the field by hydrogen gas, obtained by passing steam through red hot cylinders charged with iron turnings. The gas thus evolved was then made to pass over lime, and in this manner freed from any heavy carbonic acid gas that might adhere to it. By this method there was procured at a very moderate expense, and in the space of about four hours, a quantity of hydrogen gas sufficient to inflate a balloon 30 feet in diameter; though at first as much as 50 hours was required to produce the necessary quantity of gas.

To each war balloon there was attached a company of 30 men, under the charge of a captain, according to the report by General Baron Pelet, French Minister of War in the reign of Louis Philippe, who says also that "after five or six years' existence the Aerostation Corps was suppressed, since which time no sufficient inducement has occurred to cause that service to be reorganised in France, or to be established in foreign armies, because the perfection of balloons and their application in war render many more experiments necessary, for which the intervention of a Government is necessary."

An attempt was, however, made to revive them in the African campaign of 1830, but there was no opportunity for making use of them.

Balloons were once more used by the French in the Italian campaign of 1859. The observing work was done by civil aeronauts using a large hot-air balloon, and no results of any value were obtained.

Much better results were obtained by the Federals in the American Civil War 1860–64.

A very interesting paper on this subject was read by Captain F. Beaumont, R.E., at Chatham, on the 14th November, 1862, and has been since published in the R.E. Professional Papers for 1863, vol. xii. The following extract from his report details very clearly the system followed and the results obtained.

"There were two sizes of balloons used ; one of small size with a capacity of 13,000 cubic feet, and the other of about double this size. American apparatus.

"The larger sized balloon was the one that the Americans decidedly preferred ; it was constructed because the power of the other was found to be insufficient, and was used exclusively in place of the smaller one which it superseded.

"The balloons were made of the best and finest description of silk, double sewn and prepared with the greatest care; the summit of the balloon containing the gas valve being made of either three or four folds of cloth, to ensure sufficient strength in that part subject to the greatest strain. The varnish on which the success of the apparatus much depends, was a secret of Mr. Low's, the chief aeronaut; his balloons kept in their gas for a fortnight or more, and he attributed this to the fact of the varnish being particularly good. The network covering the bag was gathered in, in the usual manner, and ended in a series of cords attached to a ring, hanging about level with the tail of the balloon, and from this hung the wicker-work car, the ring being about level with a person's chest when standing upright in the car. Balloons.

Car.

"The string for working the valve passed through the centre of the balloon, and coming out at the tail, was loosely tied to the ring, to which were fastened the guys, three in number ; thus the car, though swayed about by the motion of the balloon, hung always nearly vertically beneath it.

"The gas generators, two in number, were nothing more than large tanks of wood, acid proof inside, and of sufficient strength to resist the expansive action of the gas ; they were provided with suitable stop-cocks for regulating the admission of the gas, and with man-hole covers for introducing the necessary materials. Generators.

CHAP. I.   MANUAL OF MILITARY BALLOONING.

"The gas used was hydrogen, and, indeed for practical purposes, all things considered, there is none other that is nearly so suitable; its low specific gravity makes it a *sine quâ non* for a military aeronaut, as independently of the ease with which it is produced, when a balloon is attached to the earth it is of the first importance that it should offer as little resistance to the air as possible, as its stability depends upon this point.

"The hydrogen was generated by using dilute sulphuric acid and iron; any old iron, such as bits of the tires of wheels, old shot broken up, &c., was used; so that it was necessary to provide only the sulphuric acid, which in large quantities is cheap, and with proper precautions very easy to carry.

Purifiers. "The gas generated passed through a leathern tube into a lime purifier, and thence in a similar manner into a second, the action of the lime simply absorbing the carbonic acid and other extraneous gases, and sending the hydrogen, quite, or very nearly pure, into the balloon. On leaving the generator its temperature was high, even the leathern pipe being so hot that the hand could hardly bear to touch it, but after passing the second purifier it was delivered, barely warm, into the balloon.

"The whole of the apparatus was so simple that nothing more remains to be said about it.

Use. "In using it the balloon is unpacked and laid in well ordered folds on a carpet spread on the ground to receive it; the tail is then placed ready for connection with the last purifier, properly charged with lime and water, and the connection by leather pipes between the purifier and the generator having been established, the latter is charged; care must be taken not to complete the communication between the last purifier and the tail of the balloon until a clear stream of hydrogen is obtained, so as to avoid getting foul air into the machine.

"Under ordinary circumstances, in three hours from the time of the machine being halted, it can be prepared for an ascent; but this, should circumstances require it, might be shortened by using two generators and making a suitable alteration in the purifying arrangement. Such alteration, however, would rarely be necessary, as the balloon, when inflated, can, unless in very windy weather, be very readily carried; 25 to 30 men lay hold of cords attached to the ring and march along, allowing the machine to rise only sufficiently to clear any obstacle that there may be in the way.

"I have frequently seen it carried thus, without the least difficulty.

"At the time I joined M'Clellan's army, it was encamped on the Pamunkey river, one march below the now celebrated White House; it was pushing its way slowly up the Peninsula, driving the Confederates before it.

"The character of this part of Virginia is much the same as

that of most parts of the agricultural districts of our own country, except that it is somewhat more undulating, and not nearly so highly cultivated. Including woodland, perhaps not half the land is under cultivation. Thus the character generally of the country is such as to render all reconnaissances, though the more desirable, very difficult to make. My first acquaintance with the balloon was made during the advance of the army; I had ridden forward from the main body and joined General Stoneman's command, then occupying for the first time, the west bank of the Chickahominy river. I found the balloon snugly ensconced in a hollow, protected from view by the hill in front, from the top of which a convenient position for an ascent was gained; the Professor's tent and those of the rest of the balloon corps were scattered round, forming a small distinct encampment.

"I received from them great civility, and was afforded every opportunity for obtaining the information I desired. It may be thought somewhat odd that such a thing as a balloon should accompany the advance of an army, but there appeared to be no difficulty in its doing so, and, of course, it was more likely to be of use there than further to the rear.

"It was employed in making continual ascents, and a daily report was sent by the principal aeronaut to M'Clellan, detailing the results of his observations; of course, in the event of anything very unusual being noticed, a special report was made.

"The observer, by continual ascents, and by noting very exactly each time the position and features of the country below him, soon knows it, as it were, by heart, and a glance is sufficient to assure him that no change has taken place in the occupation of the country.

"The balloon never got more than about a mile nearer to Richmond than when I first saw it; it may, therefore, be interesting to describe generally the position of the army, and to state what the balloon did, and what it did not do.

Topographical remarks.

"At that point the Chickahominy runs within about seven miles of Richmond, its nearest point is $4\frac{1}{4}$ miles at the village of Mechanicsville.

"It is in dry weather a sluggish stream, fordable almost at any place; but in wet weather it requires bridging, and sometimes overflowing its banks, converts the valley in which it runs into a swamp a mile wide.

"High wooded ground borders the valley of the Chickahominy on either side, one bank of the river being occupied by the Confederate army, with Richmond in its rear; the other bank being held by the main body of the Federals, who, with an army of 100,000 men, were extended over a front some 12 miles in extent, about the centre of which the balloon was stationed.

CHAP. I.   MANUAL OF MILITARY BALLOONING.

"So near to Richmond, the wished-for goal, it may be well believed that the results of the balloon ascents were looked for anxiously.

"From them were obtained the first glimpse of the Confederate capital, the capture of which, it was hoped, would virtually put an end to the war.

"Independently, though, of curiosity, most anxious inquiries were made from the observers in the balloon as to the difficulties that lay on the road to Richmond. Were there any fortifications round the place? Where were the camps, and for how many men? Were there any troops in movement near the present position? and many other questions of equal importance.

"Now, these questions were difficult to answer; and, even from the balloon, many of them could only be replied to with more or less uncertainty.

"From the balloon to the Chickahominy, as the crow flies, was about two miles; thence on to Richmond eight more.

"At the altitude of 1,000 feet, in clear weather, an effective range of vision of 10 miles could be got; thus the ground on the opposite side of Richmond could be seen; that is to say, houses and the general occupation of the land became known.

"Richmond itself was distinctly seen, and the three camps of the Confederates could be distinguished, surrounding the place.

Extent of vision.

"Looking closer, the wooded nature of the country prevented the possibility of saying whether it were occupied by troops or not; but it could be confidently asserted that no large body was in motion.

"In the same way, on seeing the camps round the place one could form a very rough estimate of the number of men they were for, but it was impossible to say whether there were men in them or not.

"Earthworks, even at a distance of eight miles, could be seen, but their character, so far off, could not be distinctly stated, though one could with certainty say whether they were of the nature of field or permanent works.

"The pickets of the enemy could be made out quite distinctly, with supports in rear, thrown forward to the banks of the stream.

"The country, from its thickly wooded character, was peculiarly unfitted for balloon reconnaissances; had it been a plain, like Lombardy, the position of any considerable body of troops would have been known; as it was, it was only possible to say that they were not in motion; this could be confidently asserted, as, though they might remain hid in the woods while stationary, in marching they must, at some time or other, come into open ground and be seen.

Hanover Courthouse.

"During the battle of Hanover Courthouse, which was the first engagement of importance before Richmond, I happened to be close to the balloon when the heavy firing began. The wind was rather high, but I was anxious to see, if possible, what was going on, and I went up with the father of the aeronaut. The balloon was, however, short of gas, and, as the wind was high, we were obliged to come down.

"I then went up by myself, the diminished weight giving increased steadiness; but it was not considered safe to go higher than 500 feet on account of the unsettled state of the weather.

"The balloon was very unsteady, so much so that it was difficult to fix my sight on any particular object; at that altitude I could see nothing of the fight.

"It turned out afterwards that the distance was, I think, over 12 miles, which, from 1,000 feet and on a clear day would, in a country of that nature, have rendered the action invisible; had the weather been such as to have allowed the balloon to remain at its usual altitude, the position of the engagement from the smoke created could have been shown, and it could have been said that no retreat had reached within a certain distance of the point of observation.

"It is quite possible, too, that with an altitude of 2,000 feet the action might have been indistinctly seen, even at the distance of 12 miles.

"At York Town, where the Federals were attacking the line of works thrown across the Peninsula, between the York and James rivers, the balloon was used continually. York Town.

"I once saw the fire of artillery directed from the balloon; this became necessary, as it was only in this way that the picket, which it was desired to dislodge, could be seen; however, I cannot say that I thought the fire of artillery was of much effect against the unseen object, not that this was the fault of the balloon, for had it not told the artillerists which way the shots were falling, their fire would have been more useless still.

"During the first two days of the heavy fighting by the left of the army before Richmond, which ended in its retreat from the Peninsula, a telegraph was taken up in the car, and the wire being placed in connection with the line to Washington, telegraphic communications were literally sent, direct from the balloon above the field of battle, to the Government. In place of this, the wires should have gone to the Commander-in-Chief's tent, or, indeed, anywhere better than to Washington, where the sole report of the state of affairs should have been received from no one but the officer in command of the army. Telegraphic communication from balloon.

"If balloons or telegraphs are to be turned into means for dividing authority, every true soldier will look upon them as evils hardly unmitigated; but this with us need not be the case, for as military machines they would be solely under the control of the Commander-in-Chief.

CHAP. I.   MANUAL OF MILITARY BALLOONING.

Opinion on value.

"General Bernard, the commanding Engineer with M'Clellan, of whom I particularly asked the question, said that he considered a balloon apparatus as decidedly a desirable thing to have with an army; but, at the same time, it was one of the first incumbrances that, if obliged to part with anything, he should leave behind.

"I myself think that it is a thing which, if properly organized and worked, may be occasionally of considerable advantage, and occasions might occur when the absence of such information as the balloon gives an opportunity of obtaining would be very bitterly felt. The observer from the balloon might, and most probably would, often enough, have nothing to report that the General did not know, but the time, on the other hand, might come when his report would contain facts, or satisfactorily confirm other information received, of such a nature that it would be invaluable. Nothing ought either to be accepted or condemned by its utility alone, but rather by its utility as compared with the cost of obtaining it; now, of the utility under certain circumstances of overlooking a tract of country from a height of 1,000 or 2,000 feet, if necessary, there can be little doubt; at the same time the cost of being able to do so is so trifling that it would appear unwise to neglect the necessary steps to secure the advantage."

Since the American Civil War, balloons were used by the French at the Siege of Paris in 1870–71.

In spite of the fact that there was at that time no balloon organization in the French Army, most valuable services were rendered by free balloons navigated almost without exception by untrained aeronauts.

Sixty-four balloons were despatched during the siege, carrying letters and despatches amounting to a weight of nearly a ton, besides a number of carrier pigeons for the return messages.

It is also well known that Gambetta left Paris in a balloon to organize an army of relief in the south of France.

No such dramatic episode has occurred since 1870 in the annals of military ballooning, but the experiences of that time have been so taken to heart by the nations of Europe, that no modern army would be considered complete without a balloon equipment.

Improvements have since taken place in the manufacture of the envelope and in the method of carrying gas; and useful results have been obtained on service, notably by the French in Tonkin in 1884–85.

Towards the end of 1884 a balloon section, under the command of Major Elsdale, R.E., accompanied Sir Charles Warren's expedition to Bechuanaland. The expedition encountered no opposition, but, so far as the transport of ballooning plant was concerned, the experiment proved entirely satisfactory. Several ascents were made near Mafeking in April, 1885, no leakage from tubes or damage to balloons having occurred during the long overland journey from Cape Town.

The balloon detachment for Bechuanaland had hardly started before it was decided to send another detachment to Suakim. Major Templer and Lieutenant Mackenzie embarked with a small detachment in February, 1885. On 26th March a 7000 cubic feet balloon was filled at Suakim, and accompanied a convoy, consisting of the Berkshire Regiment and 1,200 camels, to Sir John McNeill's zareba. On two previous occasions convoys to the zareba had met with considerable opposition from the Arabs; but on this occasion not a shot was fired nor a single camel's load lost. The balloon was in the air for eight hours, and produced an excellent moral effect, the Arabs being observed to be dispersing in all directions. On the march to Tamai, although the balloon was only up for a short time, the same moral effect was produced. Later on a small balloon of 4,500 cubic feet capacity was sent to Atao and Tambuk. A light weight Arab, weighing about six stone, was employed to make ascents. His observations proved generally correct, and as this small balloon was kept at work for more than ten days, and its supply of gas had only taken five camels to bring it out, it must be considered a very satisfactory performance, the more so as the hills in the neighbourhood of Tambuk averaged 2,000 feet in height, necessitating correspondingly high balloon ascents. The fact that a balloon detachment was at the same time being employed in South Africa handicapped the Suakim detachment considerably, the available personnel and plant being thus rendered far more limited than was desirable.

Chapter II.

MANUFACTURE AND DETAILS OF BALLOONING PLANT.

Comparison of field generator and compressed tube systems.

For military ballooning, in which lightness of equipment and mobility are such essential points, the principle of carrying gas compressed in steel tubes has been adopted by most of the nations of Europe.

Field generators were used in the American war of 1860—1864, but the manufacture of steel has vastly improved since then, and although such an apparatus may be employed again in the future, it will probably be in siege warfare only, as the superior mobility of wagons containing gas compressed in tubes over those loaded with zinc or iron and acid makes them far more suitable for employment with an army in the field.

It may be assumed, as a rough rule, that 1 ton of zinc and 1 ton of sulphuric acid are required to produce 10,000 cubic feet of hydrogen, or a fill for one balloon; in the field, allowing for leakage, waste of material, &c., it would not be safe to count on producing each fill at less than about $2\frac{1}{2}$ tons of generating materials.

An approximate weight of 4 tons of water for diluting the sulphuric acid makes the weight of materials necessary in the field generating system from 6 to 7 tons exclusive of the weight of generators.

The tube at present in use in the British service weighs 75 lbs., and contains under full pressure about 120 cubic feet of hydrogen, and at peace pressure about 100 cubic feet.

As a rule, at peace manœuvres it is found that a fill will not require more than 100 tubes, representing a weight of slightly over 3 tons.

The advantages of the compressed tube system may then be stated as follows:—

(1) The weight of materials required per fill is about one-half what is necessary in the field generator system.

(2) Greater facility of transport, concentrated sulphuric acid especially being awkward material to carry on board ship or in the field.

(3) Greater rapidity of filling, the time required being from 20 minutes to half an hour as against at least three hours in the other system.

(4) The buoyancy of the hydrogen used is superior. Hydrogen compressed in tubes is, in balloon phraseology, "dead," while that produced in the field generator system is known as "live" hydrogen. The latter, unless very perfectly washed

MANUFACTURE AND DETAILS OF BALLOONING PLANT. CHAP. II.

and dried, is difficult to hold in the balloon, and is very liable to rot the envelope.

The first gas tubes in use in the British service were designed in 1882 by Captain Templer in conjunction with Lieutenants Macdonald and Trollope.*

These tubes were tested up to a pressure of 6,000 lbs. per square inch, and were intended for an effective load of 3,000 lbs. per square inch.

In 1883 the same officers designed a tube which has been in use up to the present time—1896—and has done good service both at home and abroad.

About 500 of these tubes were made in 1885-86, and have been sent by ordinary transport to Bechuanaland, Suakim, and Roorkee, without damage to the tube or loss of gas.

The following rules were got out as to the loading and testing of these tubes:—

(1) Any tube to be passed for service, or retained as serviceable on retesting, must pass a water test of 2,500 lbs.

(2) Tubes which pass a test of 2,800 lbs. and upwards will be marked first class tubes.

(3) All other tubes will be marked second class tubes.

(4) The first class tubes will have a ring of black paint 3 inches wide on the conical part near the valve.

(5) The second class tubes will be painted white all over.

(6) All tubes will be loaded to 1,500 lbs. for peace practice, and this load will be adhered to in war time for second class tubes. *Working load.*

(7) The load of first class tubes will be increased to 1,800 lbs. for war. This allows a margin of 50 per cent. over the working load for safety, plus an allowance of 100 lbs. to allow for any ordinary increase of temperature. If the increase of temperature is likely to be extraordinary, the tubes must be loaded under the 1,800 lbs. to allow for it in accordance with the rules relating to temperature below, it being always provided that the maximum calculated working pressure at the highest expected temperature shall be two-thirds of the water test stamped on the tube.

(8) Oxygen will never be loaded above 1,500 lbs., in any tube, and its pressure will be tested with a special oxygen gauge.

A difference or rise of temperature of 100° F. corresponds to an increased pressure of one-fifth. *Rules as to temperature.*

Or an increase of 25° corresponds to an increased pressure of $\frac{1}{20}$ or 5 per cent.

Thus to allow in loading for a probable rise of temperature of 25° allow about—

* Formula for calculating dimensions and strength of wrought iron and steel tubes:—

Thickness in inches = $\dfrac{\text{pressure in lbs. per sq. in.}}{\text{safe tensile strength of material}} \times \dfrac{\text{internal diameter}}{2}$

(M.B.) C

	6 atmospheres less in all tubes,
	7 „ for 30°,
or	8 „ for 35°.

The only subsequent alteration made in these rules was that no tubes were to be loaded at a higher pressure than 102 atmospheres (or 1,500 lbs. per square inch) under any circumstances.

This was done for the sake of uniformity, in order to avoid possible accidents while compressing; an accident being more likely to happen if greater pressures were used for some tubes than for others.

The compressing station is now so arranged that, at a slight excess on the 1,500-lbs. pressure, the compressing pumps themselves begin to pull up, the safety valves open, the revolutions become slower, and cease altogether at about 1,700 lbs. per square inch, thus rendering it impossible to overpress the tubes.

The old system of obtaining extra pressure by doubling the water pressure—technically known as "topping"—has been discontinued since 1892 as being dangerous.

Security in working with these high-pressure tubes is obtained by a periodical hydraulic test of 2,700 lbs. per square inch.

A margin of safety of at least 1,000 lbs. per square inch is thus ensured.

The following test was made of one of these tubes in 1895, that is to say, after about 10 years' use. Tube No. 223 (originally marked as a first class tube) under water pressure took a permanent set at 3,300 lbs. per square inch, and took an absolute pressure of 4,000 lbs. to burst it. The burst consisted only of an opening of the weld at the centre of the tube for about 8 inches in length.

Experiments have since been made to improve on this tube, the direction in which improvement has been sought being reduced weight compatible with safety.

In 1892 a pattern of tube was recommended, and an order for 100 was given to the Projectile Company to carry out by their process.

Some of these will no doubt meet ballooning requirements, but difficulties of manufacture appear to have arisen, and it has been found necessary to reduce the estimated pressure.

The following extract from the register of tubes kept at the School of Ballooning will serve to show the difference between the 1883 pattern tube and that recommended in 1892:—

SECTIONS OF VALVE FOR TUBE.

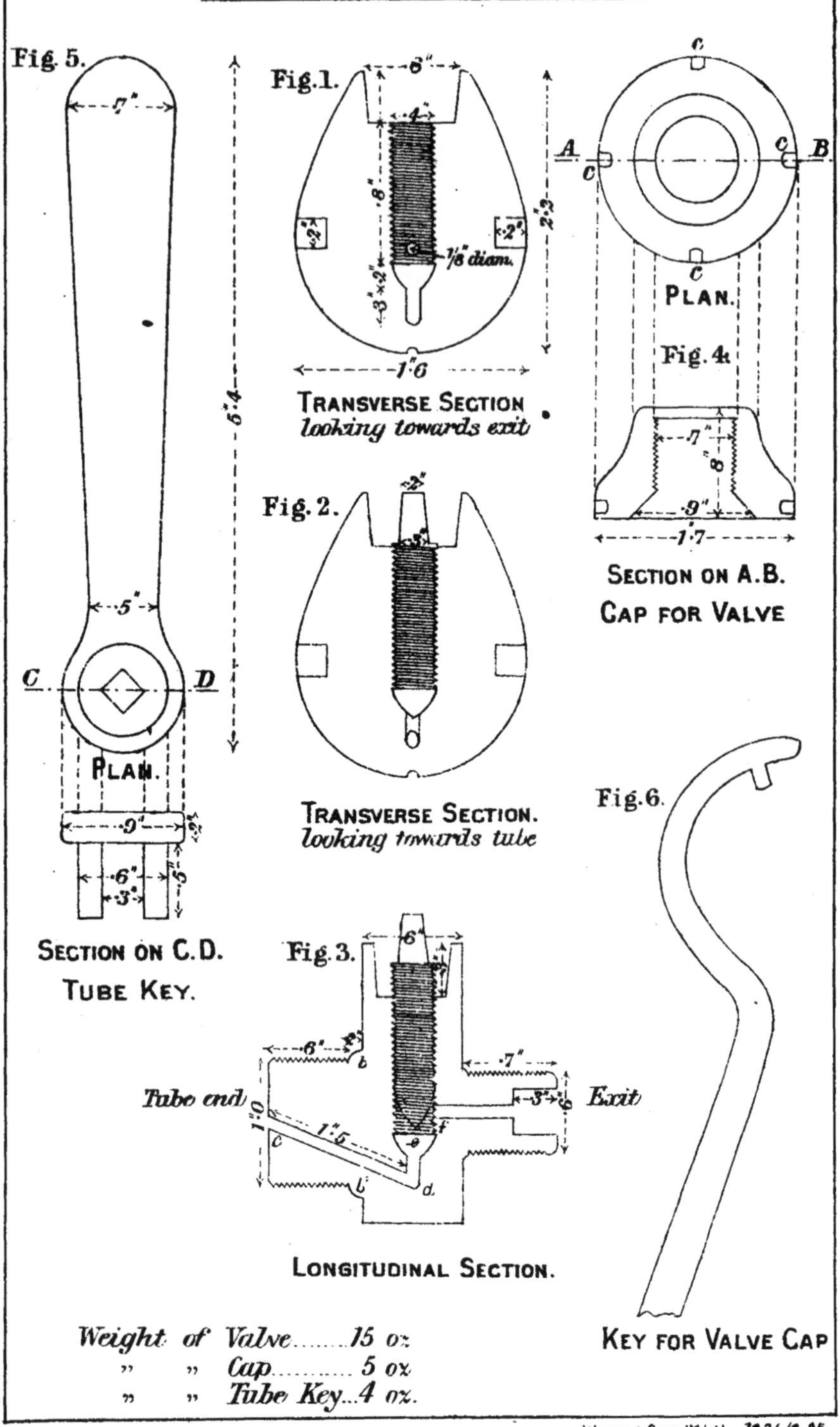

Wyman & Sons, Ltd, Lith, 7921/12.95.

MANUFACTURE AND DETAILS OF BALLOONING PLANT.   CHAP. II.

Tube No.	Length.	Diameter.	Weight.	Contents.	Balloon hydraulic test.	Working load.
223, 1883 Pattern	8 feet	5¼″	75 lbs.	1·2 cubic feet	3000 lbs.	1500 lbs. per square inch, or 102 atmospheres.
20, 1892 Pattern	8 feet	8″	100·75 lbs	2·4 cubic feet	2500 ,,	1500 lbs. per square inch, or 102 atmospheres.*

The advantage of the later pattern of tube is seen by comparing the amount of hydrogen carried per lb. weight in both cases.

In the first tube the contents are about 1·2 cubic feet at 102 atmospheres = 122·4 cubic feet of hydrogen for 75 lbs. weight of tube = 1·6 cubic feet of hydrogen per lb.

In the second tube the contents are about 2·4 cubic feet at 102 atmospheres = 244·8 cubic feet of hydrogen for 100·75 lbs. weight of tube = 2·4 cubic feet of hydrogen per lb.

In the first case one load of hydrogen (10,000 cubic feet) is carried in 82 tubes weighing 6150 lbs., or about 2·8 tons, as against 41 tubes weighing 4,131 lbs., or about 1·8 tons in the latter case.

No doubt a tube closely resembling this 1892 pattern will shortly be introduced into the service.

The fact of having obtained a steel tube which will safely stand a working pressure of 1,500 lbs. to the square inch is no good, unless the tube is provided with a valve which will stand the same pressure, and which can be so regulated as to allow the gas to be run out at any required rate. *Valve for gas tube.*

The valve at present in use has given excellent results, and has stood the test of long use and foreign service successfully; it is cast of Delta metal, and weighs on an average 15 oz.

Figs. 1, 2, and 3 show two cross sections of this valve and one longitudinal section. It may be said to consist of three parts, the inlet channel, the valve proper, and the exit channel.

(1.) The inlet channel consists of a cylinder 0·6 inch long and 1 inch diameter, with a screw thread turned on the surface; this screws into the nozzle of the tube, and is sealed by the shoulders b, b' (Fig. 3). A circular hole, ⅛ inch diameter, runs downwards, from the centre of the tube end to the valve proper. *Inlet channel.*

(2.) The valve proper is 0·7 inch thick, and of section as shown in Figs 1 and 2. A circular hole is bored vertically downwards through its centre, with a section, as in Fig. (1), and the inlet channel runs in at the bottom. *The valve proper.*

A plug with a screw thread cut on it screws into the hole,

* Since reduced to 60 atmospheres or 882 lbs. per square inch.
(M.B.)   c 2

and terminates in a cone which exactly fits the conical seating. This plug is turned down from a steel rod, and is tempered in water at a dull red heat.

The exit channel.
(3.) The exit channel consists of a cylinder 0·7 inch long and 0·6 inch diameter, with a screw thread on the outside, and a hole $\frac{1}{8}$ inch diameter bored through its centre, and communicating with the hole in the valve proper.

Action.
When the plug in the valve proper is screwed tight home, the cone at the end fits tightly into the conical seating; and the gas passing along the inlet channel (c, d), Fig. 3, is checked at the point (e); as the plug is unscrewed the gas rises, until it reaches f, the entrance to the exit channel, and passing through this channel is liberated.

Fig. 2 shows the valve shut with the plug screwed home, Fig. 3 the valve wide open, and the plug clear of the exit orifice.

The greatest height of the valve is 2·2 inches, and the greatest breadth 1·6 inch.

Tube key.
In order to facilitate opening the valve, the top of the screw plug is squared, and a wrench called the tube key (Fig. 5), is provided and carried with the tubes. It is forged from cast steel, and a square hole to fit the top of the plug punched in it.

Valve cap.
To prevent the exit channel of the valve getting broken, or the orifice stopped up the cap, Fig. 4, is made, which screws on to the exit channel. It is made of gun metal, and weighs 5 oz.

Key for caps.
There are four small holes round it, c, c, c, c, in Fig. 4, for tightening it up, if required, and a wrench (Fig. 6) must be used for screwing or unscrewing them. When transporting tubes, great care should be taken that these caps are attached, for if the screw thread on the exit channel of the valve is even slightly damaged, the valve is spoilt, as it cannot be properly connected up to the wagons or pumps.

Old pattern valve.
An older pattern of tube valve is made of phosphor bronze, and is of exactly the same construction, with the exception that the shoulders on the inlet channel are wanting, and it is necessary to solder them into the tubes. This process is called "flooding" the valves, and is sometimes done even with the delta metal valves. The seating at the neck of the tube and the seating of the inlet channel of the valve should be tinned. The solder used contains 1 part of lead to 3 parts of block tin.

Loading the tubes.
The tubes are loaded in the compressing station by being connected up in batches of six by means of copper unions, with a Brotherhood's double action compression pump.

In order to prevent leakage from the valves while being loaded, the steel plugs should, before the tubes are connected up, be coated with luting composed of a mixture of beeswax and tallow.

MANUFACTURE AND DETAILS OF BALLOONING PLANT. CHAP. II.

If this is not done, there will be serious leakage round the screw threads of the plugs; besides waste of gas, this may cause serious injury to the man working in the compressing station, especially if any arsenious acid gas is present in the hydrogen.

To remove the valve from the tube if solder has been used, the neck of the tube must be heated until the solder melts, and the valve is then unscrewed. This must never be done unless absolutely necessary, as the valve is very likely to become damaged. *Removing the valve.*

The Envelope is constructed of several thicknesses of gut, spread in the first instance on long, narrow, wooden templates. The long strips so formed are called gores, and when connected up to one another form the envelope of the balloon. *The Envelope.*

Let V be the required volume of the balloon, then assuming the balloon to be spherical, *To calculate out the gore of a balloon.*

$$V = \frac{4}{3}\pi r^3$$
$$= \frac{4}{3 \times (2)^3} \times \pi \times r^3 \times (2)^3$$
$$= \frac{1}{6}\pi D^3$$
$$= \frac{1}{6\pi^2} \times \pi^3 D^3$$
$$= \frac{1}{6\pi^2} \times (2\pi r)^3$$
$$= 0.01688 \times (\text{circ.})^3$$
$$\therefore \text{Circumference} = \sqrt[3]{\frac{V}{0.01688}}.$$

The balloon is built up of as many gores as will make W (the greatest width of the gore) a convenient size.

Assuming this number to be n,

$$W = \frac{\text{circumference}}{n}.$$

In the case of the normal balloon where V = 10,000, *Calculations for the gore of a normal military balloon of 10,000 cubic feet capacity.*

$$\text{Circ.} = \sqrt[3]{\frac{10.000}{0.01683}}$$
$$= 83.9864$$
$$= 84 \text{ feet approx.}$$

A convenient width of gore in this case is 3 feet.

CHAP. II.   MANUAL OF MILITARY BALLOONING.

The balloon will thus be seen to be constructed of $\frac{84}{3}$ = 28 gores.

Next find diameter of balloon = D.

Circumference as found above = 84 feet = πD.

$$\therefore D = \frac{84}{\pi} = \frac{84}{3\cdot14159}$$

$$= 26\cdot7 \text{ feet.}$$

Describe a circle of diameter D on any convenient scale, as shown in Fig. 7, the shape of the tail being the curve of reverse curvature, the diameter at the bottom being about 1 foot.

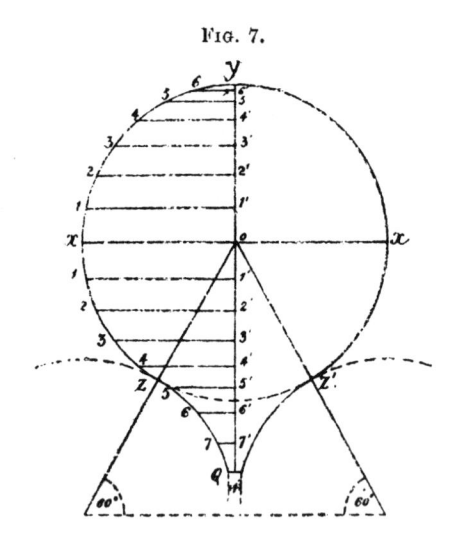

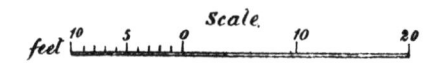

FIG. 7.

Then YXZQ = total length of gore.

From X set off round the circumference X1, 12, 23, &c., each equal to 3 feet—that length having been decided on as a convenient width of gore at its widest part—and draw 11', 22', 33', &c., parallel to XX'.

Measure each of these lines, and multiply by $\frac{\pi}{n}$, and we get the required offsets for the gore.

Thus by measurement

XO = 13·35 feet
11' = 13·2 „
22' = 12·0 „
33' = 10·6 „
44' = 8·5 „
55' = 6·0 „
66' = 3·0 „
77' = 1·5 „

Now $\dfrac{\pi}{n} = \dfrac{3\cdot 14159}{28} = 0\cdot 112$ approx.

∴ required offsets are as follows :

At 0 ...... $\dfrac{3}{2}$ .... = 1·5 feet
„ 1' ...... 13·2 × 0·112 = 1·48 „
„ 2' ...... 12·0 × 0·112 = 1·34 „
„ 3' ...... 10·6 × 0·112 = 1·19 „
„ 4' ...... 8·5 × 0·112 = 0·95 „
„ 5' ...... 6·0 × 0·112 = 0·67 „
„ 6' ...... 3·0 × 0·112 = 0·34 „
„ 7' ...... 1·5 × 0·112 = 0·17 „

Next draw a straight line ABC = YXZQ = length of gore.

Measure lengths from B each way, equal to the lengths X1, 12, &c., round the circumference, set off the calculated offsets at right angles to ABC, and join up.

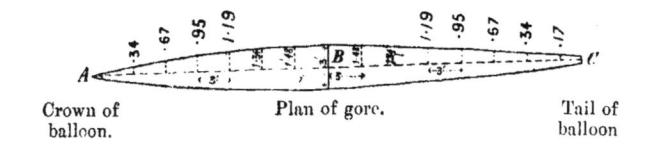

Crown of balloon.   Plan of gore.   Tail of balloon

The gore having been calculated, is next drawn out in full size on Willesden or brown paper, and handed over to the carpenters, who make a wooden gore, usually in about four pieces, which dovetail into one another. One or 1½-inch overlap must always be allowed for on one side of the gore, and if strapping is to be laid on the gore, the position of the strapping should be marked.

The following sizes of skin balloons have been made by th Balloon Department:—

CHAP. II. MANUAL OF MILITARY BALLOONING.

(1) 10,000 *Cubic Feet.*

Diameter .... 27 feet.
Great circle .... 84 ,,
No. of gores............ 28
Width of gores .... 3 feet
Weight of balloon (strapped) about 90 lbs.

This balloon is made to lift two men easily.

(2) 7,000 *Cubic Feet.*

Diameter .... 23 feet 10½ inches.
Great circle ......,.... 75 ,,
No. of gores 25
Width of gores........ 3 feet.
Weight (strapped) about 70 lbs.

This balloon lifts one man easily, two in fine weather.

(3) 4,500 *Cubic Feet.*

Only one balloon of this size has been made. Its weight is 37 lbs., and it has no top valve. It will lift one light man.

(4) 1,000 *Cubic Feet.*

This balloon and the following patterns are for signalling and photographic purposes :—

(5) 600 *Cubic Feet.*

(6) 370 *Cubic Feet.*

Weight, 5 to 10 lbs.

(7) 120 *Cubic Feet.*

Weight, 2 to 4 lbs.

(8) 2 *Cubic Feet* }
(9) 1 *Cubic Foot* } Pilot Balloons.

Valve collar. A hole is cut in the crown of the balloon to take the top valve, and a valve collar is built into the balloon. This is circular, and made of metal. Weight, 3½ lbs.

Neck hoop. At the tail of the balloon a small wooden hoop, 1 foot in diameter and about 3 inches deep, is fastened in, and from this an extra length of tail, called the petticoat, is added, about 1 foot long.

MANUFACTURE AND DETAILS OF BALLOONING PLANT.   CHAP. II.

The net ring, which clasps the valve collar, is made of flexible wire, bound with cord to prevent its cutting the upper meshes of the net. *Net ring.*

The top valve screws into the valve collar. It is actuated by pulling a cord, known as the valve line, which passes through the balloon into the car, so as to be readily accessible to the aeronaut. Weight 5¼ lbs. *Top valve.*

The net surrounds the envelope of the balloon, and distributes the weight evenly over it. *Net.*

The method of making the net is as follows:—A certain number of meshes—from 102 to 107 in the case of a normal balloon—are cast on to the ring which fits round the valve collar. This number is kept all through, but the size of the mesh is gradually increased up to the equator, to allow of the greater expanse to be covered, and is then reduced again down to the tail. The cord used is described according to its weight per 150 feet, and is spoken of as 1-lb. cord, 1⅛-lb. cord, &c. Average weight of net for 10,000 balloon from 30 to 37 lbs., made of 1-lb. cord.

The Hoop serves to connect the net and car, and is made of best seasoned ash in two pieces with long splices, the pieces being glued together and screwed up with copper screws every 4 inches. *Hoop.*

There are at present three sizes of hoops, one for the 10,000 balloon, one for the 7,000 balloon, and one which can be used for both sizes.

The dimensions, &c., are as follows:—

No. 1. Diameter from centre to centre, 2 feet 6 inches. Section, 2½ inches × 2 inches. Weight when fitted, 13½ lbs.
No. 2. Diameter, 2 feet 3 inches. Section, 1¼ inches × 1½ inches. Weight when fitted, 8¼ lbs.
No. 3. Diameter, 2 feet 4 inches. Section, 2 inches × 1¾ inch. Weight when fitted, 10 lbs.

There are three things to be provided for in fitting the hoop, viz.:— *Fittings for the hoop.*

(1) The car.
(2) The net.
(3) The captive rope.

(1) The car is attached by six ropes. Each of these ropes is passed round the hoops and made fast by an eye splice which is covered with canvas. Length of rope about 1 foot 9 inches, ending in a large toggle which fits into the car lines.

(2) The lowest meshes of the net are provided with small loops, and 107 small boxwood toggles are fastened round the hoop, and are fitted to the net.

(3) Four ropes at right angles to each other are spliced

CHAP. II.   MANUAL OF MILITARY BALLOONING.

round the hoop and brought a little below the centre, where they are all formed into a loop and spliced: the wire rope is made fast to this loop by means of a hook.

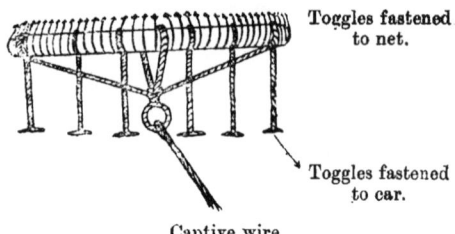

Toggles fastened to net.

Toggles fastened to car.

Captive wire rope.

Sketch showing the hoop and attachments.

Cars.   The cars are made of strong wicker work, some of the lighter sorts being strengthened with hemp rope. There are two sizes, viz.:—

(1) 10,000, car size : length, 3 feet 6 inches.
               breadth, 2 ,, 3 ,,
               depth,   2 ,, 3 ,,
               weight, 24 lbs.

Fitted with two wicker seats and six car lines, each about 3 feet long.

(2) 7,000, car size : length, 3 feet 0 inches.
              breadth, 2 ,, 0 ,,
              depth,   2 ,, 0 ,,
              weight, 20 lbs.

Fitted with one wicker seat and six car lines.

The grapnel.   The grapnel used is made of iron. It has four flukes, and is collapsible for convenience of carriage.

The head is large, in order to allow a hole, through which the rope passes, to be made. It consists of (1) centre piece, (2) four flukes, (3) four strengthening pieces, (4) a sliding piece, (5) a small clamping piece.

Weight of grapnel, 13 lbs.

The grapnel rope is $1\frac{1}{2}$ inch in circumference, is 120 feet long, and weighs 12 lbs.

Wire rope.   The following are the data regarding the captive wire rope now in use:—

Weight per 100 feet ...... 9 lbs.
Circumference............. $\frac{7}{8}$-inch.
Breaking strain .... 19 cwt.

Consists of 10 strands and an insulated core. Each strand consists of seven wires.

Wagons. The transport of the balloon section consists of six wagons, each requiring four horses.

They comprise—
- 1 balloon wagon,
- 4 tube wagons,
- 1 equipment wagon.

All these are converted from General Service wagons to avoid special fittings as much as possible.

The balloon wagon carries at the tail of the wagon a drum, on which is wound 2,200 feet of wire rope; the drum is fitted with a strong brake, and with handles at each side for winding in the rope; the telephone connections are carried from the drum to a box containing the telephones. The rope runs from the drum through a pulley, which, by means of a ball and socket joint, follows any movement of the balloon. This joint is connected to a screw shaft, which causes it to travel from one end of the drum to the other, and ensures the rope being evenly distributed. There are two boxes in the middle and front of the wagon, which carry ropes and all the necessary small stores; these boxes when closed serve as seats for the men carried on the wagon. The balloon packed in its car fits between the two boxes.

The balloon wagon in foreign equipments carries a small vertical engine, and the rope is run off or wound up by machinery; this makes the wagon very heavy, and with the present system of hauling down it is found that steam power is not necessary. The balloon wagon can carry 12 men if desired.

Each tube wagon carries 35 tubes; they are arranged longitudinally in five rows with the nozzles pointing to the rear, and are connected up to a wooden chest lined with copper, into which the gas passes, and is then carried to the balloon by connections which are put on when required. The tubes are kept steady by a light iron framework, which is unfastened when the tubes are empty and have to be unloaded.

Each tube is turned on separately, and the contents of three wagons can be run into the balloon simultaneously. The equipment wagon is a general service wagon of the usual pattern, and carries all the reserve stores, the most important being a spare balloon and a spare coil of the steel wire rope.

The following are the weights of the wagons :—

	Stripped.		Loaded.	
	tons.	cwts.	tons.	cwts.
Balloon wagon	1	5	1	10
Tube wagon	0	18	2	2
Equipment wagon	0	19	1	10

Chapter III.

GAS MANUFACTURE.

The three gases at our disposal for inflating a balloon are—

    (1) Hot air.
    (2) Coal gas.
    (3) Hydrogen.

(1) Hot air is practically out of the question. Its lifting power is very small, and a heating apparatus for keeping up the temperature inside the balloon is necessary.

This is dangerous enough in a free balloon, and for captive work the system has been tried and found impracticable.

Hot air balloons would probably only be used if the gas supply failed and it was a matter of great importance to get a balloon up even for a very short time.

(2) Coal gas has a lifting power of about 35 lbs. per 1,000 cubic feet. Its specific gravity, referred to air as unity, varies from 0·4 to 0·6. Its advantages are, first, that it is cheap and easily procurable at many places in any civilized country, and, secondly, that on account of its specific heat being greater than that of hydrogen, a balloon inflated with it is less liable to any sudden loss of equilibrium due to a change of temperature than is the case with a balloon inflated with the latter gas.

(3) Hydrogen is far more costly than coal gas, but its lifting power is from 60 to 68 lbs. per 1,000 cubic feet, and therefore for use with an army in the field, and indeed for military use generally, where economy in transport is of the highest importance, it is invaluable.

The specific gravity* of pure hydrogen referred to air $=$ 0·0692.

Practically it is found that the hydrogen as manufactured and used at the School of Ballooning seldom or never has a specific gravity of less than 0·1, or a lifting power at the normal pressure and temperature of 68·037 lbs.

Hydrogen, being the lightest form of matter, is generally regarded as the chemical unit of weight and volume, *i.e.*, one part by weight of hydrogen occupies one volume.

Preparation of hydrogen. The simplest process, chemically speaking, for preparing hydrogen in quantity consists in passing steam (H_2O) over red-hot iron.

The iron (Fe) combines with the oxygen (O) of the water

---

* The specific gravity of a gas is its weight as compared with that of an equal volume of air at the same temperature and pressure.

(H_2O) to form the black oxide of iron (Fe_3O_4) which will be found in a crystalline state upon the surface of the metal.

The process may be represented by the following formula :—

$$4H_2O + Fe_3 = Fe_3O_4 + 4H_2$$
Water.   Oxide of iron.

The weight of an atom* of iron is 56 times that of an atom of hydrogen; hence the Fe_3 in the above equation represent 56×3, or 168 parts by weight of iron.

From this can be calculated the amount of iron theoretically required in order to produce a given amount (say 10,000 cubic feet) of hydrogen.

The weight of 10,000 cubic feet of air at 30 inches and 60° F. is 767·08 lbs., and the specific gravity of pure hydrogen is 0·0692.†

Therefore the weight of 10,000 cubic feet of hydrogen at the normal pressure and temperature $= 767·08 \times 0·0692 = 53·08$ lbs.

Now, from the above equation Fe_3, or 168 parts by weight of iron, produce $4H_2$, or 8 parts by weight of hydrogen. Therefore,

1 part by weight of H is produced by $\frac{168}{8}$ parts of iron.

∴ 1 lb. H is produced by 21 lbs. Fe.
∴ 53·08 lbs. H are produced by $21 \times 53·08$ lbs. Fe.
∴ 53·08 lbs. H   ,,   by 1114·68 lbs. Fe.

In other words, about half a ton of iron would be theoretically required in order to produce 10,000 cubic feet of H at the normal pressure and temperature. As a matter of fact, considerably more than this would be required, as the black oxide of iron which forms in a crystalline state upon the surface of the metal after a while prevents it from being further acted upon by the steam.

The commonest process by which hydrogen is prepared consists in dissolving iron or zinc in a mixture of sulphuric acid and water. Zinc is the most convenient metal to employ.

It is used either in small fragments or cuttings, or as granulated zinc, prepared by melting it in a ladle and pouring it from a height of three or four feet into a tub full of water.

This is the method at present employed at the School of Ballooning.

One of the advantages of zinc is its easy fusibility—tin and lead among the ordinary metals alone surpassing it. Its melting point is below red heat, and has been estimated at 770° F.

---

* For the definition of an atom and the atomic theory, see Bloxam's "Chemistry," page 8.
† See Appendix B, Table I.

CHAP. III. MANUAL OF MILITARY BALLOONING.

The preparation of hydrogen by dissolving zinc in diluted sulphuric acid may be represented by the equation :

$$H_2SO_4 + Zn = ZnSO_4 + H_2$$
Sulphuric acid.   Sulphate of zinc.

Zn here represents one atom of zinc, which is 65[*] times the weight of an atom of hydrogen.

From this can be calculated the amount of zinc theoretically required to produce 10,000 cubic feet of hydrogen, or 53·08 lbs., at the normal temperature and pressure.

Two parts by weight of H are produced by 65 parts by weight of Zn.

∴ 1 lb. H is produced by $\frac{65}{2}$ lbs. Zn.

∴ 53·08 lbs. H are produced by $\frac{65 \times 53·08}{2}$ lbs. Zn.

∴ 53·08 lbs. H   „   „   1725·10 lbs. Zn.

In other words, about three quarters of a ton of zinc would be theoretically required in order to produce 10,000 cubic feet of H.

As a matter of fact, it may be taken that in practice about one ton of zinc is required to actually obtain this amount of hydrogen.

Iron might be used in the above method instead of zinc, the process being represented by the equation :

$$H_2SO_4 + Fe = FeSO_4 + H_2.$$
Sulphate of Iron.

From the above equation Fe, or 56 parts by weight[†] of iron, produce two parts by weight of hydrogen. Hence 1 lb. H is produced by 28 lbs. Fe. Therefore, 53·08 lbs. H is produced by 1,486 lbs., or 13 cwt. of iron are theoretically required by this method to produce 10,000 cubic feet of hydrogen.

In both the above methods the same amount of H_2SO_4 is theoretically required to produce a given amount of hydrogen.

Thus, $H_2SO_4 = 98$ parts[‡] by weight, produce 2 parts by weight of hydrogen, or 1 lb H is produced by 49 lbs. H_2SO_4.

∴ 53·08 lbs. H is produced by 2580·9 lbs. H_2SO_4,

or about 1 ton 3 cwt. of H_2SO_4 is required to produce 10,000 cubic feet of hydrogen.

A fourth method of producing hydrogen is by the electrolysis of water; this process possesses the advantage of separat-

[*] See Bloxam's "Chemistry," page 287.
[†] See Bloxam's "Chemistry," page 301.
[‡] See Bloxam's "Chemistry," page 203.

ing both H and O from the compound—water (H_2O); each of these gases can be led off to a separate gasholder, and taken off as required.

Although probably this would be the most satisfactory method of producing hydrogen for balloon purposes, yet the first cost of electrical plant, &c., has, up to the present, prevented any considerable utilisation of this method.

Lastly, hydrogen has been prepared cheaply in large quantity by heating a mixture of slaked lime, chemically known as hydrate of lime, $CaOH_2O$, and anthracite coal, which contains 90 per cent. of carbon, in an iron retort; this process is represented by the formula—

$$C + \underset{\text{Lime.}}{CaO \cdot 2H_2O} = \underset{\text{Carbonate of lime.}}{CaCO_3} + 2H_2.$$

At a red heat the water is expelled from the slaked lime; quick or anhydrous lime (CaO) combining with C and O forms $CaCO_3$, $2H_2$ being liberated.

On passing steam over the residue—

$$2H_2O + CaCO_3 = CaO \cdot 2H_2O + CO_2.$$

The process can then be repeated; hence if enough carbon be employed in the beginning, large quantities of hydrogen may be obtained by steaming and heating alternately.

The zinc and sulphuric acid method is at present employed at Aldershot for the manufacture of hydrogen.

Gas production at the School of Ballooning.

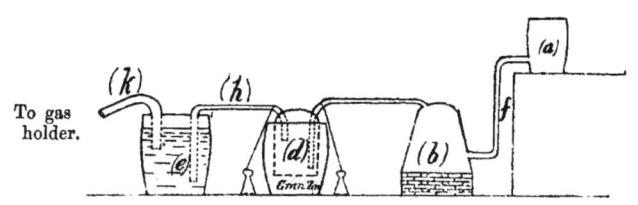

In the barrel (*a*) are placed 6 gallons of sulphuric acid (H_2SO_4) and 24 gallons of water. This mixture passes through the tube (*f*) into a copper retort (*b*) which contains 60 lbs. of granulated zinc. Hydrogen is generated, and passes into a copper vessel (*d*) shaped like a diving-bell, and sunk, by means of weights, in a barrel nearly full of water. The retort (*b*) is constantly played on by a jet of water to keep it cool. The gas is washed in (*d*), any free sulphuric acid still in the gas being taken up by some granulated zinc at the bottom of the barrel just below (*d*). It then passes through (*h*) into a cooling barrel (*e*), and thence through the pipe (*k*) to the gasholder. The acid used in this process must be free from arsenic.

If the acid supplied is made from copper pyrites, arsenious acid gas is generated, which is very injurious to the men employed in the compressing station, if there is any escape of gas. Arsenic in the solid form is also deposited after a time in the port of the tube valve, so that the tube cannot be emptied without removing the valve.

Chapter IV.
CAPTIVE WORK.

The normal rate of wind up to which it is safe to carry out captive work is 20 miles per hour. Above 30 miles per hour captive work is impossible. *Rate of wind.*

The difficulty of observing increases with the rate of the wind, considerable practice being required to observe with any good effect if the wind is over 20 miles per hour.

The length of wire-rope on the drum of the wagon is 2,200 feet. It is only in very calm weather, however, that a height of 1,000 feet can be reached, 1,500 feet being about the extreme limit than can be attained by a 10,000 cubic-foot balloon with a light man in the car in a dead calm. *Height that can be attained.*

The following is the method of laying out and filling from tube wagons. *Laying out and filling.*

Having chosen a place as much sheltered from the wind as possible, the tube wagons are formed up in line as close as possible. The balloon wagon should be close at hand in any convenient position.

First lay down a ground cloth in rear of the tube wagons to prevent the balloon from being torn by stones or twigs while being filled.

Next bring the balloon from the wagon and lay it out on the cloth. If the balloon is not in its net, lay out the net and disconnect four or five of the net toggles; two men will then get inside the net and roll out the balloon, while the remainder hold up the net and assist; three or four men will be sent to fill sand-bags.

The men then drop the net and distribute themselves evenly round the balloon; they lay hold of the envelope and draw out the folds first above and then beneath until the balloon is as shown.

Valve collar.

Tail.

(M.B.)

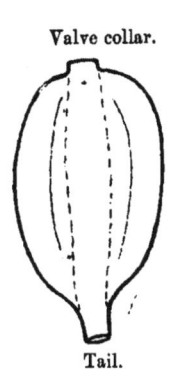

Crowning the balloon.

This consists in getting the valve into the centre, so that when the gas flows in it will raise the heaviest part, viz., the valve first.

The man at the crown pays out the valve-collar, one man on each side hauling towards the tail by means of the net, until the valve-collar is in the centre of the balloon. The folds of the envelope are then drawn out flat to form a circle round the valve-collar. The valve is then screwed in. The valve line (a ¾-in. cotton rope) should have been previously attached to the valve by the following method.

Pass the line through the eye at the base of the valve, taking two round turns with the running end on the shank of the eye, under the standing end, and through the eye the reverse way.

Next take two half-hitches on the standing end, and finish by passing the running end through a loosened strand of the standing end. The line is then rolled into a tight round ball, that will easily unwind by its own weight.

The wagons are connected up for filling the balloon by means of the hoses and zinc connections carried by them, and everything is then ready for filling. When the command is given to fill the balloon, one man at each wagon turns on the tubes gradually, and the gas flows into the balloon.

One man must be detailed to look after the connections, and one placed at the tail of the balloon.

The remainder will be distributed round the balloon to hook on the sand-bags to the net, and raise them or lower them according to the directions of the officer in charge.

When the balloon is filled the tubes are turned off, the hose disconnected, and the sand-bags lowered down to the hoop.

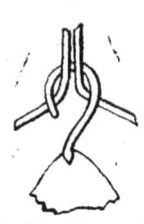

Method of hooking on Sand-bags.

The car is then toggled on, and the wire-rope connected to the hoop for a captive ascent.

Balloon Drill.

Fall in.

The Balloon Section, consisting of two N.C.O.'s and eight files, will fall in in two ranks ten paces in front of the wagons. The balloon wagon will be formed up on the right of the tube wagons.

CAPTIVE WORK. CHAP. IV.

The men number off alternately front and rear rank. The odd numbers are thus in the front rank, and the even numbers in the rear rank. *For balloon drill, number.*

        Odd numbers   ..   Balloon men.
        2, 4, 6, 8   ..   ..   Tube men.
        10, 12, 14, 16   ..   Sandbag men.

The balloon men double to the balloon wagon, the tube and sandbag men to the tube wagons—two men to each tube wagon, commencing from the right. *On the wagons, prepare to mount.*

The Co.-Serjt.-Major will go to the balloon wagon, and the senior serjeant to the tube wagon on the left.

Extra men will double to the balloon wagon.

The N.C.O.'s and men mount on their respective wagons. *Mount.*

The dismounting should be done some way from the place selected for filling. *From the wagons, prepare to dismount.*

The N.C.O.'s and men dismount, and fall in their previous order in front of the wagons.

On reaching the place for filling, the dismounted men will ground arms about 20 yards away, while the tube wagons are being formed up as close together as possible. *Dismount.*

The balloon men double to the off side, and the sandbag men to the near side of the balloon wagon. The tube men double to the tube wagons, No. 2 taking the right hand, No. 4 the centre, and No. 6 the left hand wagon. *Prepare to lay out.*

Nos. 1, 3, 5, and 7 take the car; Nos. 9, 11, 13, and 15 the ground cloth from the wagon, and carry them to the rear of the centre tube wagon. These numbers then spread the cloth, unpack the balloon, and roll her out. *Lay out.*

No. 1 stands at the tail, No. 3 at the crown, taking hold of the net; Nos. 5, 7, and 9 are on the left of No. 1, and Nos. 11, 13, and 15 on his right. These numbers will then overhaul the envelope, drawing out the folds, until they assume the position shown on page 33. Nos. 2, 4, and 6 take out the skin hoses from their wagons and connect to a 3-way filler which No. 8 brings up. Nos. 2, 4, and 6 then get their keys ready, and stand on their wagons ready to turn on. No. 8 connects the 3-way filler to an 8-in. zinc piece, and the latter to the petticoat of the balloon. He then stands by the 3-way piece, not leaving it till the filling is completed.

Nos. 10, 12, 14, and 16 take the sandbags from the wagon, fill them and place them round the edge of the cloth.

The balloon men take hold of the net. *Prepare to crown.*

No. 3 slacks away the net, and Nos. 5 and 11 haul in. The other numbers draw the upper folds of the envelope towards the tail, and the under folds outwards, so as to form a circle with the valve-collar in the centre. *Crown.*

As soon as this circle is formed, the commander gives the word, "Down," when the net will be arranged evenly on the *Down.*

(M.B.)   D 2

CHAP. IV.   MANUAL OF MILITARY BALLOONING.

envelope, and the men will kneel down, holding the envelope to the ground.

Valve up. Prepare to turn in. No. 3 takes the valve and line, which is properly rolled, places the latter inside the balloon, and screws the valve home. No. 5 assists No. 3, and looks after the luting for the valve.

Turn in. The tube men turn in, commencing with the top layer of tubes.

The balloon men are kneeling round the balloon, holding the envelope down; No. 1 is at the tail, No. 8 at the 3-way piece.

The sandbag men under the serjeant see to the fitting up of the car, and that all stores to be taken up are ready in the car.

The balloon and sandbag men then let up the balloon as directed.

Turn off. No. 1 disconnects from balloon, sees that the valve-line is clear, and fastens up the petticoat. The sandbag men bring up and toggle on the car. No. 3 hooks on the wire cable. Nos. 10 and 12 secure the grapnel rope. Nos. 15 and 16 stand by the brake on the balloon wagon.

The remainder of the men, holding the net in their left hands, then remove the sandbags with their right hands.

The observers take their places in the car.

The ballast is adjusted according to the lift of the balloon, which is then ready for an ascent.

Return stores. The balloon, tube, and sandbag men return stores to their respective wagons.

When the balloon is filled from tubes on the ground, the tubes are usually arranged in three heaps containing 40 tubes in each heap. Each heap of tubes should be piled in the form of a truncated pyramid containing 10 tubes in the lower row and six in the upper one. Each tube is then separately connected with a forty-filler by a piece of india-rubber tubing. See sketch below.

When the wire rope is made fast to the balloon and everything is ready, the man at the brake will ease off gently, and allow the balloon to ascend. He should never allow the drum to get out of hand, but in case of accident a man may be placed on top of the wagon with a handspike which he will use as a brake if ordered.

When the balloon has reached the required altitude for observing purposes the brake will be lashed down, the wagon as a rule being stationary. If there is much wind the wheels should be lashed together, or else picketted down.

It should be remembered that the wind currents are frequently local, and if the balloon is at first riding roughly she should be let up or down to try and run into a quieter current. When the wagon is in motion the great aim should be to make the balloon travel as easily as possible. When going against the wind the wagon must move slowly, as otherwise the

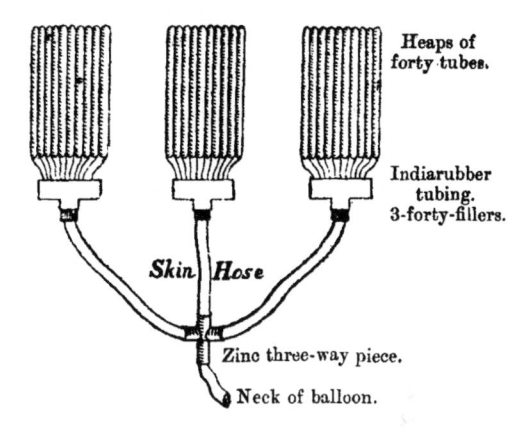

```
                    Heaps of
                    forty tubes.


                    Indiarubber
                    tubing.
                    3-forty-fillers.

          Skin  Hose


          Zinc three-way piece.
         Neck of balloon.
```

balloon is beaten down by the wind and a halt has to be made to let it recover itself, entailing a loss of time and a certain amount of discomfort to the aeronaut. The aeronaut should, if possible, be informed beforehand of any proposed movement of the wagon or the balloon.

That of the latter should as a rule be regulated by the aeronaut himself. The telephone is not always audible, especially when the wagon is in motion, so a code of signals for such commands as Haul Down (D), Halt (H), Let the Balloon up (U), &c., should be arranged.

One of the most important things to be watched in captive work is the expansion of gas through the tail. This is especially necessary at the commencement of practice, the loss of heat due to sudden expansion of the compressed gas, causing it to flow into the balloon in a very cold state. *Expansion.*

The hydrogen will not, therefore, be fully expanded for some little time after filling.

Now, in captive work, economy of gas is very important, so whenever possible the petticoat should be tied up by a slip-knot. One of the aeronauts should therefore sit in the hoop and pay special attention to the expansion. If the tail becomes in the least distended, the petticoat should be at once opened to allow of free expansion.

The more open the country the better for transporting a balloon about. Towns, villages, and woods must be avoided, and if too great a detour is required to go round by road, the wire rope must be taken off and the balloon taken across country by hand on two cotton ropes. *Obstacles.*

If, however, the line of march and the direction of the wind are almost coincident, the balloon could be weighted so as to be in equilibrium 200 or 300 feet up. When the wagon had been *Going free over an obstacle.*

CHAP. IV. MANUAL OF MILITARY BALLOONING.

disconnected it would be driven sharply through the town or wood, and the men with it would place themselves so that the the balloon would float over them after passing the obstacle.

The balloon would then be allowed to go free, and the aeronaut would trail a rope for the men to catch hold of.

Trees.

Tall trees by the side of the road are awkward if the wind is from either flank; if they occur continually the balloon must be worked at the leeward side of them. The following is the method of clearing a single tree or house which the rope is about to foul.

The dotted line is the first position of the wire rope, see sketch below. Unless steps are taken to prevent it, the rope will foul the tree. Take a cotton rope, a, b, and make it fast with a stopper hitch to the wire rope, and let the latter run out

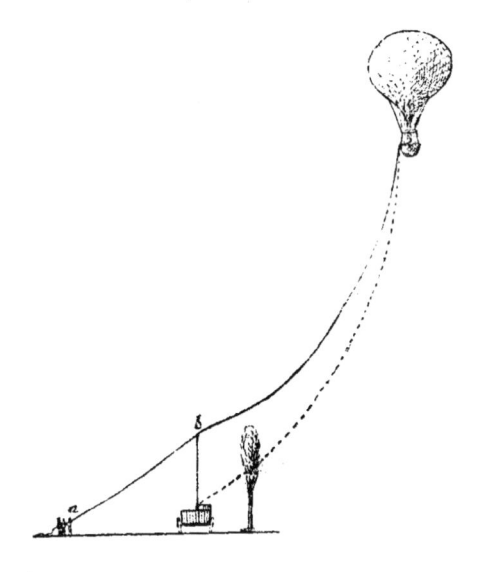

to a convenient distance; now send the cotton rope with three or four men to the side of the road furthest from the obstacle, and a little judicious working will bring the ropes into the position shown by the continuous line. The wagon can now be driven past the obstacle and the auxiliary rope removed.

Telegraph wires.

To pass these the balloon must be taken off the wagon and put on two cotton ropes.

If the wind is in the direction of the line of march, one rope is coiled up in the car.

The balloon is then let up on the other rope and allowed to float over the telegraph wires.

CAPTIVE WORK. CHAP. IV.

When well over on the lee side of the wires, the aeronaut throws out his rope from the car, and when this has been seized by the men on the ground, he hauls up the windward rope, which in turn he drops on the lee side when clear of the wires. See Fig. 1.

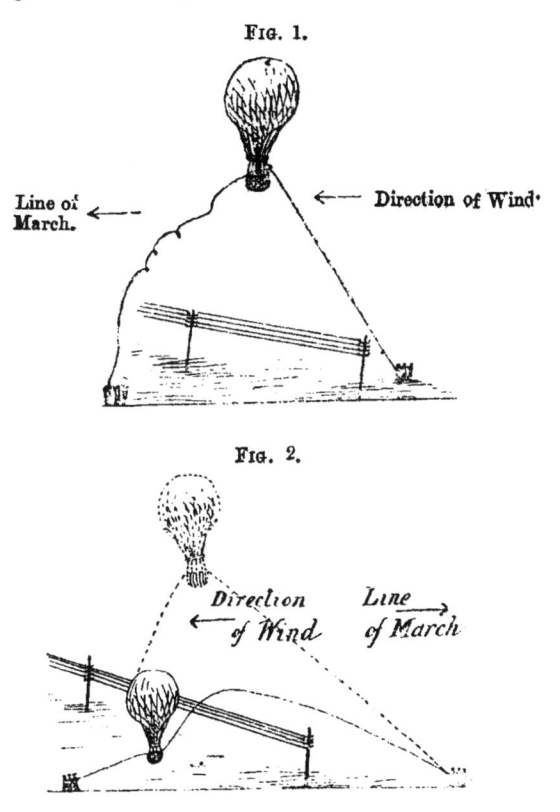

FIG. 1.

Line of March. ← ← Direction of Wind.

FIG. 2.

Direction of Wind ← ← Line of March →

When the wind is against the balloon the rope must be thrown over the wires before the balloon ascends, and the balloon towed over by it as shown in Fig. 2.

To bring a balloon down by winding up at the wagon is a very long and rather dangerous job in a strong wind and should never be tried where it can be avoided. The usual way is as follows. **Hauling down.**

A snatch block is fastened by a rope to the centre of a 6-ft. handspike; the snatch block is then put on to the captive rope, three men take hold of each end of the handspike and walk

CHAP. IV.   MANUAL OF MILITARY BALLOONING.

down towards the balloon which is thus brought down without bringing much strain either on it or on the ropes. A clear space cannot always be got, but on reaching any obstruction a spare snatch block can be made fast to a tree or post, the rope passed through it, and the men can then change direction.

Another method is to keep the snatch block and handspike stationary, and drive the wagon in an opposite direction. The two methods can also be used together, and the balloon brought down very quickly.

Buoys.

There are some occasions when the available lift of the balloon is not sufficient to attain the required height.

For instance, in South Africa, the starting point was 5000 ft. above the level of the sea, and the lifting power of the gas correspondingly diminished.

In such cases a balloon may be made to rise much higher by using balloons of the 120′ or 370′ cub. ft. capacity at intervals along the captive rope to buoy up the line as shown in the diagram.

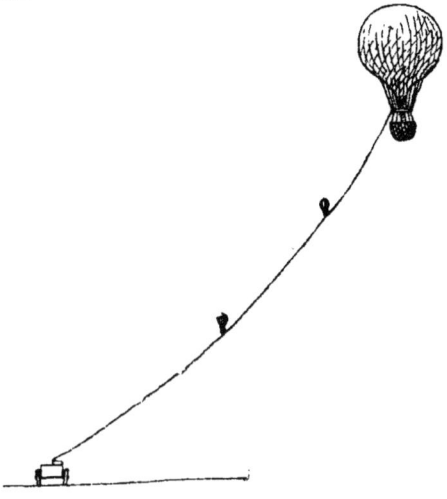

Communications.

The communication between the balloon and wagon can be kept up either by telephone or by means of message-bags, small flags being only suitable for a code of signals such as the one described earlier in this chapter. The most convenient method of sending messages and plans from the balloon is by means of the message bag—a small canvas bag which is weighted and provided with a split ring at one corner. The message is put in the bag, the split ring made to enclose the wire cable, when the weight of the bag makes it run down the rope to the ground.

CAPTIVE WORK. CHAP. IV.

The telephone is the best means of communicating messages, if the wagon is stationary, and there is not too much noise of firing near the wagon. The following general results have been arrived at with reference to the best method of communication between the car and wagon.

(1) The wire cable at present in use cannot always be trusted to keep its insulation perfect.

(2) The telephone is often unnecessary during field manœuvres, the frequent hauling down to avoid fire making communication from the ground to the car unnecessary.

(3) Message bags transmit reports from the car as rapidly as the telephone, especially when there is any noise from firing, or from horses hooked into the wagon.

(4) Where a fixed station is used, telephone communication from the ground to the car would often be very useful. In this case it would be easy to use an insulated return wire, the chance of its being broken being very slight when the wagon is not in motion.

On the conclusion of a day's captive work, the balloon should be bagged down for the night in the most sheltered spot available. *Bagging down.*

If the night is calm, and there is a chance of the balloon being required for night observation, it may be sufficient to keep the balloon weighting the car with sand-bags.

If, however, there is any wind the car must be untoggled, and the balloon properly bagged down. This is done as follows:—

About 6 feet from the ground small cord loops are fastened round the net all round the balloon, and through the loops a cotton rope is passed. To this rope are hung full sand-bags till the balloon is sufficiently weighted to prevent the chance of beating about.

The ground cloth should be placed under the balloon before bagging down, and to prevent the envelope from being chafed, some softer cloths should be laid over it.

A guard should be mounted over the balloon at night, with orders to empty the balloon at discretion if the wind gets up to such an extent as to threaten to tear the envelope by beating it about on the ground.

When doing captive work, the wire rope should always, as far as possible, be connected electrically with the earth. *Precautions against lightning.*

This should be done at all times, but more especially in thundery weather.

To insulate the wagon by placing it on a thick ground cloth should be specially avoided.

Chapter V.

FREE RUNS.

Section (1).—*Theoretical Considerations.*

The management of a free balloon can only be thoroughly learned after considerable practice under varied atmospheric conditions, but much useful knowledge can be obtained by an attentive consideration of the theory of the subject.

Principle of Archimedes. The principle of Archimedes, that *a body immersed in a liquid loses a part of its weight equal to the weight of the displaced liquid*, is equally true of bodies immersed in air.

If the body in question is heavier than air, it will remain on the surface of the earth; if it is as heavy as air, its weight will counterbalance the buoyancy, and it will just float in the atmosphere; if it is lighter than air, the body will rise until it reaches a layer of the atmosphere of a density equal to its own.

Boyle's law. The second law, which it is most important to bear in mind in all ballooning calculations, is that known as Boyle's Law; it refers to the compressibility of gases, and may be stated as follows:—

The temperature remaining the same, the volume of a given quantity of gas varies inversely as the pressure which it bears, the weight of a given volume varying directly as the pressure.

This law was at one time thought to be of universal application; subsequent experiments have, however, shown that carbonic acid, sulphuretted hydrogen, ammonia, and cyanogen are more compressible than air; and that hydrogen, which has the same compressibility as air up to 15 atmospheres, is then less compressible.

For ballooning calculations, however, the law may be taken as absolute.

The third law, which we will next consider, refers to the expansion in volume of gases due to rise of temperature, and may be thus expressed:—

All gases expand very nearly alike, $\frac{1}{490}$ of their volume = (0.00202) *for* $1°$ *Fahr., or* $\frac{1}{273} = (0.00366)$ *for* $1°$ *Centigrade.*

As an illustration of the first law, suppose a 10,000 cubic feet balloon inflated with hydrogen of specific gravity 0·1 (a good sample of gas) at 30 inches and 60° F.

The weight of 10,000 cubic feet of dry air under these conditions .. .. = 767·08 lbs.
(See App. B, Table I.)
The weight of H of sp. gr. 0·1 .. = 76·71 „

Now, from the law of Archimedes, that a body (in this case hydrogen) immersed in a liquid (in this case air) loses a part of its weight equal to the weight of the displaced liquid, the loss of weight of the hydrogen = 76·71 — 767·08 = algebraically speaking —690·37.

In other words, its weight has now become a min*u*s quantity, having been converted into ascensional force or lifting power.

Assuming the weight of the envelope, net
car, &c., to be .. .. 200 lbs.
And weight of two occupants to be .. 320 ,,

Total .. 520 ,,

The amount of ballast necessary to produce equilibrium will be 690·37 — 520 lbs. = 170·37 lbs.

As an illustration of Boyle's law, let the problem now be to find the amount of ballast which it will be necessary to expend in order to reach an altitude at which the barometric reading is 25 inches.

Assuming the temperature to remain unchanged, this will mean an altitude of about 5,000 feet. (See App. B, Table VI.)

The pressure of 25 inches being $\frac{25}{30}$, or $\frac{5}{6}$ of that at the ground level, a volume of air occupying 10,000 cubic feet at ground level will at the lower pressure occupy a space of $\frac{6 \text{ cubic feet} \times 10,000}{5}$ = 12,000 cubic feet, and the weight of 10,000 cubic feet of air at 25 inches will be $\frac{5}{6}$ of its weight at 30 inches, that is to say, $\frac{5}{6}$ of 767·08 lbs., which equals 639·23 lbs. (See App. B, Table IV.)

Similarly, the weight of 10,000 cubic feet of hydrogen of sp. gr. 0·1 at 25 inches pressure = $\frac{5}{6}$ of 76·71 = 63·92 lbs.

The lifting power of the gas at this elevation will thus be 639·23 —63·92 = 575·31 lbs.

We have already seen that the weight of the balloon and occupants amounts to 520 lbs., so that 575·31 —520 = 55·31 lbs. will be all the ballast now remaining in the car.

We also saw that 170·37 lbs. of ballast produced equilibrium at the ground level, so that 170·37 —55·31 = 115·06 lbs. of ballast must have been expended in order to reach an altitude of 5,000 feet, the temperature being assumed to remain at 60° Fahr. throughout.

From the above example, a general formula of ballast output can be deduced, temperature being assumed to be constant, and the balloon to be full and in equilibrium.

Formula of ballast output.

Let C = cubic capacity, in feet, of the balloon.
Let L = lifting power, in lbs., of 1 cubic foot of gas at the lower station.

CHAP. V.   MANUAL OF MILITARY BALLOONING.

Let P = barometric pressure, in inches, at lower station.
Let p =   „   „   upper station.

Then, if x be the number of pounds of ballast required to be expended in order to rise from a position at which the barometric pressure is P inches to a higher point at which the pressure is p inches :

$$x = CL \times \left(1 - \frac{p}{P}\right).$$

Next, let us consider what takes place in the case of a falling balloon.

Assume balloon full of gas of sp. gr. 0·1, and in equilibrium at 20 baro. and 60° Fahr. (an approximate height of 11,000 feet, see App. B, Table VI).

Weight of air (see App. B, Table IV)   = 511·4 lbs.
Weight of H .. .. = 51·1 „
                                   Lift .. = 460·3 „

We may suppose this weight to be made up as follows :—

Weight of envelope, net, &c.   ..   ..   220·0 lbs.
Weight of 1 aeronaut   ..   ..   ..   160·0 „
Ballast   ..   ..   80·3 „
                              Total   .. 460·3 „

Suppose now that the valve is opened, and 100 cubic feet of gas allowed to escape—

Weight of 10,000 cubic feet of gas .. = 51·14 lbs.
∴ Weight of 100 cubic feet of gas   .. = 0·511 „
∴ Remaining weight of H ..   .. = 50·6 „

Similarly, weight of displaced air (*i.e.*, 9,900 cubic feet)

= 511·4 − 5·1 = 506·3 lbs.
∴ Lift = 506·3 − 50·6 = 455·7 lbs.

But the balloon was assumed to be in equilibrium when lifting 460·3 lbs., so that equilibrium is now destroyed and the balloon will descend, 460·3 − 455·7 = 4·6 lbs., being the excess weight which the gas cannot lift.

This amount is known as the Rupture of Equilibrium.

Next consider the balloon at 25 baro. and 60° Fahr.

The gas will continue to contract as the balloon falls, so that at this lower elevation none will have escaped from the balloon.

Now, at 20 inches pressure the gas occupied a space of 9,900 cubic feet, therefore, by Boyle's Law, at 25 inches pressure it will occupy a space of $\frac{20}{25} \times 9,900 = 7920$ cubic feet.

The weight of 10,000 cubic feet of air at this temperature and pressure $= \frac{25}{20}$ of $511\cdot4 = 639\cdot25$. (See App. B, Table IV)

∴ The weight of 10,000 cubic feet of gas of sp. gr. $0\cdot1 = 63\cdot925$

∴ ,, ,, 1 cubic foot ,, ,, ,, $\frac{63\cdot925}{10,000}$

∴ ,, ,, 7,920 cubic feet ,, ,, $\frac{63\cdot925 \times 7920}{10,000}$

$= 50\cdot628$

∴ The weight of the displaced air $= 506\cdot286$

∴ The lift $= 455\cdot7$,

which is the same as it was before, and the balloon will continue to descend. In the same way it can be shown that at 30 inches pressure and 60° Fahr. the lift will still be 455·7 lbs., the weight of the gas being still 50·6 lbs., and the weight of the displaced air 506·3 lbs.

Suppose now on reaching the ground that ballast to the amount of 4·6 lbs. (the rupture of equilibrium) be thrown out.

The ballast now remaining in the car will amount to 80·3 —4·6 lbs. = 75·7 lbs., and the total weight to be raised will be 460·3 —4·6 = 455·7 lbs.

But we have just seen that the lift of the slack balloon has been exactly this amount throughout its descending course; the balloon will therefore once more be in equilibrium.

It is necessary, however, to explain at this point that the term "in equilibrium," when used in ballooning, refers almost exclusively to a full balloon which has reached its greatest elevation.

Theoretically, a slack balloon may be in equilibrium, but this equilibrium will be found in practice to be of a most unstable character; this is no doubt to a great extent due to the flapping of the tail, which causes a continual alteration in the cubic capacity of the envelope, and, consequently, in the weight of the displaced air.

No doubt a balloon may occasionally be made to travel some little way while in this unstable condition of equilibrium, but the aeronaut will have to work with his valve line in one hand and a bag of ballast in the other, a mode of progression which, it is needless to point out, is neither economical of gas nor of ballast, and which tends very soon to bring the run to an end.

Let us now consider the effect of throwing out another pound of ballast; this will cause the balloon to rise, the rupture of equilibrium being 1, the total weight to be raised being 454·7 lbs.

It will be readily seen, as the lift of the balloon has been found to be 455·7 lbs. up to 20 inches pressure, that the balloon will rise up to this height, at which point the gas will have expanded so as to occupy 9,900 cubic feet, *i.e.*, the volume of the gas after opening the valve before descending.

It will, however, rise beyond this point (the rupture of equilibrium being still 1) until fully expanded, at which point, the weight of gas being still 50·6 lbs., and the weight of the displaced air being 506·3 lbs., the lift will still be 455·7 lbs.

From this point upwards the balloon will lose gas by expansion through the tail, until the weight of gas remaining in the balloon, and the weight of the displaced air, are diminished to such a point that the difference between them is reduced from 455·7 lbs. to 454·7 lbs., at which moment the balloon will be in equilibrium.

In order to ascertain the altitude the balloon will eventually reach, let us first find its altitude when fully expanded.

At this point the weight of the displaced air = 506·3 lbs.; now, from App. B, Table IV, the weight of 10,000 cubic feet of air at 30 inches and 60° Fahr. = 767·1 lbs.

Therefore the barometric pressure now reached (*i.e.*, when the balloon is just full) = $\frac{506·3}{767·1} \times 30 = 19·8$ inches. From the formula of ballast output—

$$x = CL \times \left(1 - \frac{p}{P}\right),$$

the ultimate point that will be reached can now be found.

The problem in this case is to find p, x being 1, *i.e*, the rupture of equilibrium, and P being 19·8 inches.

At this pressure the lifting power of 10,000 cubic feet of gas of sp. gr. 1 = $\frac{19·8}{30}$ of 690·4 = 455·7, as we have already seen above.

Substituting in the above equation—

$$1 = 455·7 \times \left(1 - \frac{p}{19·8}\right)$$

$$= 455·7 - \frac{455·7 \times p}{19·8}$$

$$\therefore p = \frac{454·7 \times 19·8}{455·7} = 19·73 \text{ inches.}$$

From App. B, Table VI, it will be seen that this pressure indicates an approximate height of 11,400 feet, or about 400 feet above the original position of equilibrium.

We have now followed the course of a balloon from a height of 11,000 feet to the ground, and its upward course back again to an altitude of 11,400 feet, and have seen that the rupture of equilibrium both when falling and rising remained a constant quantity.

Hence follows the general law that *when a slack balloon rises or falls, its rupture of equilibrium remains constant, and its velocity of ascent or descent is also a constant quantity.*

On the other hand when a balloon of constant volume (*i.e.*, a full balloon) rises and loses gas by expansion through the tail, its lift diminishes progressively and with it the velocity.

From the foregoing theoretical considerations, the following laws may be deduced, the temperature being assumed to be constant :—

(1) A balloon, when once it starts falling, will continue to fall until it reaches the ground.

(2) If the fall is checked by throwing out ballast and the balloon rises again, it will find its equilibrium at a point above its original position.

(3) If the fall is checked and the balloon remains in equilibrium at a low level, its equilibrium will be very unstable.

(4) The higher a full balloon rises, the more gas it loses from expansion.

Theoretically we have seen that the velocity of ascent and descent of a slack balloon is constant; as a matter of fact, its velocity of descent is appreciably accelerated as the balloon approaches the ground.

This is partly due to the flapping of the envelope, and consequent variations of capacity; and also, if the tail is left open, to the inrush of air into the balloon from below.

To obviate this latter cause of acceleration, the man in the hoop should hold the petticoat in his hand while the balloon is descending.

If a dangerously rapid descent is being made, and there is not sufficient ballast in the car to reduce the shock of landing within safe limits, the attachment of the petticoat to the hoop may be cast off, and the balloon allowed to parachute; this method should not, however, be adopted if it can possibly be avoided, especially in windy weather, as the balloon will beat about on the ground while being emptied much more than it would otherwise do, and will very likely be badly torn.

Another point in which practice does not at first sight appear to square with theory is seen in the behaviour of a balloon on a sudden fall of temperature.

As a matter of fact, whether this fall of temperature is caused by the balloon entering a cloud, or by the sun being obscured by a cloud, the balloon will always fall. In the first instance, the added weight of moisture on the envelope would,

CHAP. V.   MANUAL OF MILITARY BALLOONING.

no doubt, tend to cause a drop, but in the latter case no such explanation can be given.

If the contraction of the air and hydrogen, due to the fall in temperature, happened instantaneously, theoretically by the third law given above, the balloon should remain in equilibrium (unstable equilibrium, however, as the balloon would now be slack); but, as a matter of fact, this alteration in density does not take place either instantaneously or contemporaneously, the hydrogen assuming its new density first, before the surrounding air has fully contracted.

This accounts for an immediate loss of lift, the density of the hydrogen having increased in a higher ratio than that of the surrounding air.

This result is partly due to the fact that the hydrogen being confined by the envelope is heated when in the sun's rays to a higher temperature than the surrounding air, thus causing the balloon to travel above its proper altitude. It is also, no doubt, partly due to the fact that the specific heat of hydrogen is less than that of air.

Comparing equal volumes of these substances, their specific heats are given below:—

$$\text{Air} \ldots \ldots \ldots \ldots 1\cdot000$$
$$\text{Hydrogen} \ldots \ldots 0\cdot993$$

The specific heat of air is thus seen to be greater than that of H; in other words, it requires more heat to raise the temperature of air through 1° than it does to raise the temperature of an equal volume of hydrogen to the same extent; conversely, if equal volumes of air and H be allowed to cool, the air will take a slightly longer time to cool than the hydrogen.

For purposes of free runs, it will thus be seen that the best gas to use would be one whose specific heat is the same as that of air, as in this case changes of temperature would affect the two substances equally, and the balloon would be much steadier under varied conditions of temperature than is the case when it is filled with hydrogen.

The skin envelope, however, no doubt affects the result not only by confining the gas, but also by absorption and radiation of heat, so it would be better still to use a gas whose specific heat is greater than that of air, for instance, coal gas, whose specific heat is probably always greater than that of air.

Thus, if two balloons of equal lifting power, one containing coal gas and the other hydrogen, were started simultaneously on a free run, it would be safe to predict that the coal gas balloon would do the longest distance.

FREE RUNS.

Section 2.—Instructions for the Conduct of a Free Run.

The following stores must be carried:—   Preliminary Stores.

In the car—

    Grapnel and grapnel rope.
    Aneroid barometer.
* Fahrenheit thermometer.
    Map of the country.
    Balloon cloth.
    Old newspaper.
* Bradshaw's Railway Guide.
* Manual of Military Ballooning.
    Knife and sheath.
    Free run book.
    Coil of cord.

On the person—

Watch.	Note book.
Money.	Pencils.
* Compass.	* Flask.
* Field glasses.	* Sandwiches.

If the balloon has not much lift, those stores marked with an asterisk (*) may be omitted in order to allow their equivalent in weight to be taken in ballast.

The following operations are to be carried out personally by the officer or N.C.O. in charge of the balloon, and also, for the sake of practice, by any other person under instruction who is present.

(1) Test the valve to see that it opens and shuts correctly, and that the valve line is properly hung and of convenient length.

(2) Overhaul the grapnel rope, examine its attachment to the hoop, see that it is coiled up on the side of the car so that it will run out freely when required, and make certain that the conical plug at the ground end of the rope is securely spliced in.

(3) See that the grapnel opens easily, that it is correctly made fast to the side of the car, and that the clip fits tightly.

(4) Set the barometer.

(5) Carefully check the stores in the list already given.

(6) Examine the ballast, and see that it is as free from stones or lumps as circumstances permit.

(M.B.)   E

CHAP. V.   MANUAL OF MILITARY BALLOONING.

(7) See that the petticoat is open, so as to allow of free expansion of gas.

(8) Send up one or two pilot balloons, take their bearing with a prismatic compass, plot the direction of their course in pencil on the map, and note carefully any change of direction in the higher currents of air.

If possible, authority for the run must be obtained beforehand, and the senior should provide himself with an order for journeys of officers, when not proceeding with troops under route (A.F.O., 1799), and one or two forms for booking clerks (A.F.O., 18,200) for use on the return journey.

Ballasting up. The operation of ballasting up the balloon must be done by an experienced man, and it is more convenient if he is not one of the aeronauts.

The aeronauts having taken their places in the car, and the stores having been placed in their proper positions, ballast is placed in the car until the balloon is in perfect equilibrium on the ground, so that if any ballast is taken out the balloon will rise. The weight of ballast so taken out is called the "lift" of the balloon, and the balloon is described as going away with a 5-lb. lift or whatever the weight may be.

The exact amount of lift required for a balloon to reach a given height can be arrived at from the formula of ballast output given in the preceding section, viz., $x = CL \times \left(1 - \frac{p}{P}\right)$; but the conditions of the ground, atmosphere, gas, &c., vary so greatly that by far the best guide is practical experience.

The principles governing the question of what is the best height for the balloon to attain at the start will be considered later, but it is evident that it must clear any obstacles immediately in front of it. When a balloon is started on its journey from a clear open space, the determination of the correct lift is an easy matter; but when the start has to take place from an enclosed ground of small area, there are one or two difficulties which complicate the question, and the following causes have to be dealt with :—

The true force and direction of the wind are masked by the surrounding buildings, and a balloon which seems to have sufficient lift at starting may be driven down as soon as it feels the true wind, or be driven against a higher obstacle which appeared to be considerably out of the line of flight.

Another effect of the wind in a sheltered place is to give the balloon what is called "false lift;" this is usually caused by little eddies of wind, which get under the balloon and give it an upward movement which is almost immediately lost. False lift may also be caused by the man in charge giving the car a jerk up as he lets go, instead of allowing the balloon to float away quite naturally.

From these considerations it is imperative that a margin of

safety should always be allowed. On a fine day, with no obstacles in the line of flight, the velocity of the wind being between 5 and 15 miles per hour, a lift of not more than 2 lbs. should be given.

The chief reasons for starting with as little lift as possible being—

(1) To ensure a prolonged run. This will be explained later on.

(2) To avoid a rapid ascent, which causes expansion to take place so quickly that in order to prevent overshooting the position of equilibrium, and then commencing to fall at once, the expansion may have to be eased by opening the valve. This is not conducive to economy of gas, and must of course materially shorten the run.

(3) That there is always a possibility of the envelope bursting (especially if an old one) owing to the tension placed on it by the expansion of the gas taking place too rapidly.

Accidents of this nature have been known to occur with cambric or silk balloons when the material has become old and rotten.

When there is an experienced man on the ground to start the balloon the procedure is as follows:—Having ballasted up and taken out the amount of ballast he considers necessary, he orders every one but himself to let go, and, keeping both hands on the car, lets the balloon go down wind. The force he has to exert to keep the balloon down tells him what the lift is like, and if he judges it to be sufficient he lets go, and warns the aeronaut that he is free; if, however, it appears too much or too little, he orders the men to bring the balloon back, readjusts the ballast, and tries again until he is satisfied. *The start.*

When the aeronaut prefers to regulate the start himself, he grips the hand of a man who walks down wind with the balloon, and judges from the pressure he has to exert whether the lift is correct or not.

When starting, the aeronaut in charge should be standing up in the car, facing the way the balloon is going, with a bag of ballast on the seat in front of him, and the valve line close to his hand, ready to at once check or regulate an ascent or descent if required.

In the normal military balloon of 10,000 cubic feet capacity, two men can usually be carried on a free run, their duties being arranged as follows:—The more experienced aeronaut is in charge of the balloon; he alone decides when to ascend and descend, manages the valve and ballast, and regulates the arrangements for landing; his companion's duties are more those of a clerk, and he keeps the records in the "Free Run Book," and keeps a careful eye on the barometer. These duties may of course be varied when the purpose of a run is *Duties of the aeronauts.*

(M.B.) E 2

CHAP. V. MANUAL OF MILITARY BALLOONING.

to train men to manage the balloon. It will ordinarily be found convenient for the aeronaut in charge to have the car to himself, and for his assistant to sit in the hoop.

Management of the balloon during the run.

If the balloon falls, ballast has to be thrown out to check it. On the other hand, if the balloon rises too rapidly, the expansion of gas must be eased by the valve. The life of the balloon, *i.e.*, the duration of the free run, is over either when the ballast is expended or when there is not sufficient gas to produce the necessary lifting power. It follows, therefore, that ballast must be used sparingly and at the moment when its effect is greatest, and that recourse must be had to the valve line as seldom as possible. In other words, in the judicious use of valve and ballast lies the secret of a successful run.

It will be convenient now to recapitulate the laws of practical ballooning given in the last chapter.

Laws of practical ballooning.

(1) A balloon, when once it starts falling, will continue to fall until it reaches the ground.

(2) If the fall is checked by throwing out ballast, and the balloon rises again, it will find its equilibrium at a point above its original position.

(3) If the fall is checked, and the balloon remains in equilibrium at a lower level, its equilibrium will be very unstable.

(4) The higher a full balloon rises, the more gas it loses from expansion.

It follows that the height at which the balloon will travel during the run can never be counted on to be lower than the height to which it rises at the start, and that if it goes to an unnecessary height at first the balloon loses by expansion an amount of gas which would otherwise have been saved. On the other hand, if the balloon is kept too low at first, and commences to fall, there is so short a time in which to check it, that an unnecessary amount of ballast may have to be expended to avoid a collision with the ground; and taking these two reasons into consideration, the best height for the first position of equilibrium of the balloon is between 500 and 1,500 feet. The balloon will remain in this position until the conditions are altered either by the aeronaut using his valve or ballast, or by a change in the temperature of the surrounding atmosphere.

If the temperature falls, the volume of gas contracts, and the balloon falls, and, as a general rule, this fall can only be checked by throwing out ballast. The initial velocity of the descent is usually very slight, and a handful of ballast thrown out directly the barometer indicates a fall will check the balloon and restore equilibrium.

If the fall is allowed to continue for some time, the velocity becomes much accelerated, and it is a much harder matter to check it without expending too much ballast, thus causing the balloon to rise again to an altitude considerably greater than that of the first position of equilibrium.

The most common mistake in throwing out ballast is in not giving time for it to take effect. It must be remembered that the balloon is not brought to a stop immediately, and no extra ballast should be thrown out if the velocity of descent seems to be diminishing. By throwing out small pieces of newspaper and comparing their fall with that of the balloon, a very good guide to the velocity of descent is obtainable.

There are some cases in which it may be advisable not to check a fall at once, but to let the balloon run for some time. Suppose, for instance, that a small cloud in the upper currents is passing between the balloon and the sun; the temperature is lowered, and the balloon falls. What is the result of throwing out ballast at once? The balloon is brought into equilibrium to suit the fall in temperature: in a minute or two the sun shines out again, and the balloon rises to a new position of equilibrium; but if it is allowed to fall until the cloud has passed, the fall will probably be checked without the use of ballast.

If the country which the balloon is passing is fairly open, the grapnel rope, instead of being coiled up on the side of the car as at starting, may be allowed to hang down below the car; it then fulfils two duties. The barometer only indicates the height above the starting point, but the rope will show when the balloon is close to the ground by its end trailing; and if the fall continues, the weight of that portion of the rope which is trailing is to a great extent taken off the balloon, and the effect is the same as if ballast were thrown out. There is a third way of avoiding loss of ballast, but it is only practicable in open country and calm weather; in this case the balloon may be allowed to fall until it actually strikes the ground, the weight of the car and its contents being taken off momentarily, the balloon again tends to rise.

A balloon which has been driven down to the ground by a heavy shower may remain there until the rain is over, and then rise again; but it is obvious that this could only be tried on a very calm day. If the temperature rises, the balloon will rise to a new position of equilibrium, and, unless there is any special reason for remaining low down, no steps should be taken to check the ascent. An exception to this is when the balloon is likely to rise through the clouds; here the balloon passes out of a damp, cold atmosphere into the full heat of the sun's rays, which cause it to rise at a great pace, losing a lot of gas from expansion; and on falling again into the colder atmosphere below the clouds, the fall becomes very rapid and hard to check.

Unless it is of the utmost importance to continue a free run, the landing must be made while there is still a good supply of ballast in hand.

The landing.

CHAP. V.   MANUAL OF MILITARY BALLOONING.

Suppose x the height expressed in thousands of feet that the balloon must fall, the following table shows the amount of ballast in lbs. that should be kept for the descent and landing :—

In fine weather	$3x + 20$
In cloudy or foggy weather	$5x + 20$
With snow or rain	$8x + 20$

Roughly speaking, a bag of ballast of the size in use with military balloons, may be taken to weigh 20 lbs. Thus, in fine weather, for a descent from 4,000 feet, 32 lbs., or $1\frac{1}{2}$ bags, would be required; in cloudy weather, 40 lbs., or 2 bags, and, in rainy weather, 52 lbs., or $2\frac{1}{2}$ bags. The following conditions are necessary for a good landing place.

(1) Proximity to a good road and railway station.
(2) Shelter for the balloon from the force of the wind.
(3) Clear ground in front.
(4) Good holding place for the grapnel.
(5) Probability of obtaining assistance.

These are self-evident, but it is not always possible to choose a place uniting all these conditions, and No. 2 is the most important. The best place to come down is on the lee side of a wood, in a sheltered hollow, or on the reverse slope of a hill, and the close proximity of buildings should be avoided. When the aeronaut in charge has decided that it is time to make a landing, the balloon should be brought down fairly close to the ground, and be steadied as much as possible; the grapnel rope is let down, the grapnel opened ready for use, the barometer and all other small stores having been carefully secured.

Having selected a good landing place, the balloon is so manœuvred that the grapnel rope touches the ground in front of the hedge or other obstacle in which it is desired that the grapnel should catch, and the grapnel is allowed to slide down the rope.

As soon as the grapnel catches, the balloon is beaten down to the ground, and the occupants of the car must be prepared for a shock; they should be sitting on the seats or side of the car, holding on to the car lines or hoop, and taking care that their legs and arms are slightly bent, not stiff.

If the grapnel has taken firm hold, the aeronaut in charge must open the valve to its full extent and keep it so, never letting go the valve line until the balloon is brought to rest; in most cases the balloon will rise and fall two or three times before the lift is destroyed from loss of gas.

If the descent has not been accurately gauged, and there is no chance of the grapnel catching as required, it should not be let go, but some of the reserve ballast thrown out and a new landing place chosen; when once the grapnel has been let go, the aeronauts are committed to a landing (unless they have a large supply of ballast), as it takes a long time to recover it.

If the grapnel does not take hold firmly at once, it must depend entirely on the nature of the ground in front, and amount of ballast available, whether to get rid of the gas as quickly as possible and let the balloon drag, or to keep all the gas and throw out ballast to clear an obstacle in front.

On no account should either of the occupants quit the car voluntarily until the balloon is at a standstill, unless both have made a definite arrangement to that effect in a case of great danger.

A useful precaution is to tie a small lashing round the clip of the grapnel to prevent all chance of its being jerked out by the shock.

Packing up. When the balloon is sufficiently empty to prevent its rising with one in the car, one of the aeronauts may get out, and with such help as is available, haul down the balloon on its side to leeward till he can reach the valve, which he will unscrew and take out to allow of the free passage of the gas; the rest of the gas is then forced out of the balloon as far as possible. Next disconnect the car and grapnel rope, fold up the balloon lengthways, and then roll it up, commencing at the valve collar, and forcing out any gas or air still remaining in the balloon through the tail.

Lastly, roll the balloon up in the balloon cloth, place it in the car, collect the small stores, valve, and grapnel, and get to the nearest station as soon as possible.

If the balloon cloth has not been brought, an old sheet can often be bought at a cottage, and a cart for transport should be arranged for as soon as possible.

Record of the run. It has already been stated that one aeronaut is told off to keep the observations during the run; the final report is compiled from them, and these observations, or a copy of them, are included in the report.

The free run book mentioned on page 49 as one of the stores to be carried in the car, is, strictly speaking, a leather case containing a supply of three different forms, supplied at the School of Ballooning, on which are to be recorded the observations made in the air.

Form 1 acts as a cover for the others. On the outside page are printed various headings comprising all the necessary data. The inside pages are left blank for general remarks.

Form 2 is divided into six columns, headed respectively Number of Observation, Time, Height, Temperature, Place, and Remarks.

Form 3 is a chart on which is to be marked the height of the balloon at various times corresponding with those entered

CHAP. V.   MANUAL OF MILITARY BALLOONING.

in Form 2, so as to give a graphic representation of the vertical path of the balloon during the run. The horizontal scale is a time scale, the divisions and subdivisions reading periods of hours, quarter hours, and five minutes from the start; the vertical scale is one of height, the divisions and subdivisions reading distances of 1,000 feet, 500 feet, and 100 feet above the starting level.* Spaces are left below for putting down the actual time and the position of the balloon.

Before starting, the officer who is to record should lash his barometer and watch* in a position where he can see them easily without moving his position in the car; he should also fill in, as far as possible, the headings on the front of Form 1.

The usual balloon map for free run work is the Railway Map of Great Britain; its scale is 3 miles to the inch, and the railways are very clearly marked by thick red lines, tunnels are shown by a break in the line, and stations by a red square. On a very calm day, when the distance travelled is not likely to be great, a large scale map may be taken.

When the balloon starts, the time of start is recorded on Form 1, entered on Form 2 as observation No. 1, and the space headed "Actual Time" in Form 3 completed.

From this time until the completion of the run, observations are taken about every 10 minutes, entered in Form 2, the height marked on Form 3 against the corresponding time, and joined up to the height of the preceding observation. Whenever the balloon commences or finishes rising or falling, or passes over or near a railway, river, or other easily identified place, an observation must be taken and recorded. As long as the direction of the balloon remains constant and the balloon is in sight of the earth, there is seldom much difficulty in identifying the course on the map; if, however, the currents vary at different heights, it is necessary to carefully watch the places over which the balloon passes. The balloon often revolves slowly, and it is difficult to notice a change of direction by trusting solely to the compass. When the run has been in progress for a period of from a quarter of an hour to half an hour, the distance travelled, if the position is known, should be measured off the map, so as to determine the approximate speed. Knowing this, if the balloon gets above the clouds for a time, its position is more easy to fix when the earth is once more visible. There may not be time to record the last two or three observations taken just before landing; these should be booked, and all three Forms completed as soon as possible after landing to avoid mistakes. On some occasions it may happen that no map is available, or that the course of the balloon runs out of the map. In this case, small sketches of any noticeable points which are in or near the track of the balloon, should be made in the column of "Remarks" in Form 2, or on the blank

* A wristlet watch is very convenient for ballooning.

FREE RUNS.   CHAP. V.

pages in the Free Run Book, with a view to identifying the positions on a map afterwards.

In addition to the three forms already mentioned, another chart (Form 4) is required for the report. This is a height and distance chart, and the vertical track of the balloon may vary considerably from that in Form 3, if the strength of the wind during the run was variable. There is no printed form for this chart, which has to be drawn on return to quarters; it is most convenient to execute it on a sheet of cartridge paper. The scale of the plan may be any one that is convenient, preferably that of the balloon map, or half that scale if the run is a long one. It is obvious that in both of the charts the vertical path of the balloon is a distorted one, as it is impossible to make the vertical and horizontal scales the same. A covering letter in the usual form should be written, forwarding the report to the Instructor in Ballooning.

The time chart below is given as a reduced specimen of Form 3. It represents graphically the longest run accomplished up to the present by two aeronauts in a balloon of 10,000 cubic feet capacity.

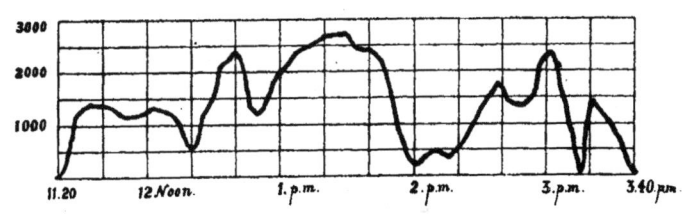

Time Chart (Form 3), showing vertical height of balloon "Tampion," 10,000 cubic feet capacity, at intervals of 20 minutes.

Aeronauts, { Captain H. B. Jones, R.E.
{ Lieutenant H. C. Prichard, Northampton Regiment.

Date of free run, 3rd November, 1894. Duration of run, 4 hours 20 mins. Distance travelled, 152 miles (from Aldershot to Louth). Average speed, 35 miles per hour.

It will be found useful to keep a record of all expenses incurred in connection with the run on a blank page in the "Free Run Book" to facilitate the making out of claims for recovering the amount. At present the sum of one pound (£1) is allowed for expenses incurred in packing up the balloon and transporting it to Aldershot; this is recovered on a Contractor's Bill Form; the fare for the return journey is also recoverable on an ordinary Travelling Claim Form, but personal allowance is not chargeable.

At present it seems a moot point with the various railway companies whether the balloon is allowed to go as personal luggage or not; in the latter case it should be forwarded by goods train, but in any case the valve and other small stores should be taken in the carriage.

A telegram, giving time and place of descent, should be sent to the Instructor in Ballooning as soon as possible after landing.

Chapter VI.

ON THE EMPLOYMENT OF BALLOONS IN WAR.

It should always be borne in mind that, apart from the question of free ascents, a balloon with its equipment simply exists as a portable observatory to assist the G.O.C. in carrying out any reconnaissance or observing work he may require. *Employment of captive balloons.*

The principal limitation to the use of a captive balloon is inability to work under ordinary circumstances with a wind blowing over 20 miles per hour. A wind of 30 miles per hour is probably the absolute limit at which a captive balloon could be worked; but the probability of getting useful reports under these conditions would be very slight. *Limitation of use due to wind.*

In England it may be assumed that captive work would be possible for three days out of five in the summer months, and for two days out of five during the winter.

The following are the most likely uses to which a captive balloon would be put so as to be of service in the field:—

(1) Reconnaissance of a defensive position occupied by an enemy, especially with a view to obtaining information as to the position and the strength of the enemy's reserves.

(2) Observation of an enemy's bivouac or camp by day or night.

By day, the approximate numbers of the enemy might be estimated by noting the number of bivouacs or camps of units. Various useful information, such as the hour of striking camp, &c., and the enemy's movements generally might be observed.

By night the number of bivouac or camp fires of a hostile force close at hand might be counted. Again, the presence of an enemy's bivouac or camp at a considerable distance might be ascertained by aid of a captive balloon, and the direction taken by compass bearing or by noting the exact hour at which any particular star was immediately above it.

It may be noted here that a knowledge of the elements of astronomy is most necessary in carrying out balloon observations at night.

(3) Observations to assist a G.O.C. in concentrating his forces, and to keep him in touch with the various columns of his own force when advancing on a broad front, before coming into contact with the enemy.

(4) Observations to assist artillery in correcting their aim when firing at unseen targets, *e.g.*, when firing at a battery on the reverse slope of a hill, or when firing at a sunken escarp wall.

(5) Observations of the enemy's movements in the interior of a besieged town. This was successfully done by the Federals

at the siege of Richmond; the concentration of the Confederates before making sorties was observed, and their intentions frustrated.

(6) Observations to ascertain the position and number of the enemy's works and batteries.

This applies equally to the attack or defence.

Photography might be usefully employed in observations of this nature.

(7) A balloon might be used as a central signal station from which a G.O.C. could communicate orders to his force.

On a calm day, when an altitude of 1,000 feet might be attained with the normal balloon, the distance of the horizon on a plain would be 38 miles. At an altitude of 500 feet the distance of the horizon would be 27 miles. A collapsing drum for signalling on the Morse system would probably be the best means of carrying this out.

(8) Observations of the enemy's movements during the course of an action, especially as regards the detaching of any portion of their forces to march round or threaten a flank. It is not perhaps very often that observations of the enemy's dispositions immediately in front would be of much practical value, although circumstances may always arise under which a General might reasonably expect assistance in framing his dispositions by the use of a captive balloon. To obtain early information of any outflanking movement, and thus give time for a General to guard against any possible attack on either flank, would be the most likely rôle for a balloon during an action.

Nos. (1), (2), (3), (7), and (8) of the above list have reference to the use of balloons in field manœuvres, and as the existing balloon equipment in the British service is specially designed for mobility, they probably represent the most useful rôle for them on service.

Nos. (4), (5), and (6), on the other hand, refer to siege warfare; and as mobility of equipment would not be so essential in this case, either for the attack or defence, use would probably be made of balloons of larger capacity than the normal military balloon of 10,000 cubic feet.

The advantages to be gained by the use of larger balloons are greater lifting power, and consequently greater elevation. A cheaper gas than hydrogen—for instance, coal-gas—might also be used.

This would be specially advantageous for the defence, as local gasworks could be made use of to inflate the balloons.

The most valuable example of the use of captive balloons at sieges was afforded at the siege of Richmond in 1862. A long extract from Captain Beaumont's report on the balloon operations before Richmond is given in the first chapter of this manual.

The general conclusions that may be drawn from his report,

and from the experiences of subsequent observers, are that on a clear day with a light wind, when an altitude of at least 1,000 feet would be attainable, movements of large forces, positions of camps, earthworks, &c., might be observed up to an extreme radius of about eight miles.

It may be safely assumed that under favourable conditions no movement of large bodies of men—such as a brigade—could take place within a radius of five miles of a captive balloon without being immediately detected.

As to the actual distance from the enemy's guns at which a balloon can be manœuvred while being fired at, we have no modern war data to go upon.

Many experiments have been made on this point both in this country and on the Continent, some favourable to the guns and others to the balloon. The great means of defence at the disposal of the balloon consists in continual change of altitude. When this manœuvre is well executed, the balloon being allowed to run up or being rapidly hauled down at the sight of the flash from the gun, the gunners have found considerable difficulty in getting the correct elevation. A stationary balloon, however, is by no means a difficult mark for field guns to hit certainly up to 4,000 yards, and probably at longer ranges. When under artillery fire, therefore, the balloon should always be moved about, if possible both horizontally and vertically.

Assuming this to be done, short ascents could no doubt be attempted to obtain some definite and important information at a distance of 3,000 yards or even less from the enemy's field guns; but 2 miles (3,520 yards), as laid down by Major Elsdale in a paper read before the Aldershot Military Society "On Military Balloons" in 1889, should be looked upon as the normal minimum distance from an enemy's battery at which it is possible to observe from a captive balloon.

In all cases, whenever possible, there should be a military telegraph station at the foot of the captive rope in communication with the G.O.C.

Under certain circumstances, the direction of wind being favourable, a charge of gun-cotton might no doubt be floated out over an enemy's work, and dropped into it. *Dropping high explosives from balloons.*

The charge would be provided with a percussion fuze or it would be fired by means of a dynamo-electric exploder in the car.

Such an operation would have to be carried out in the dusk of the evening or early dawn.

The experience of the siege of Paris in 1870–71 is sufficient answer to those who doubt the possible utility of free balloons in war time. The regular service of despatches from the besieged town, to say nothing of the departure of Gambetta from Paris by balloon to organize the army in the south of France sufficiently proves the military utility of free balloons. *Employment of free balloons.*

CHAP. VI. MANUAL OF MILITARY BALLOONING.

In the case of a town with a considerable *enceinte*, such as Paris, it would be quite possible to send despatches into as well as out of the town; but the rough-and-ready organization of the French balloons in 1870 no doubt prevented, supposing it had been considered desirable, such an attempt from being made.

Circumstances might also arise under which a useful reconnaissance might be made either over a besieged town or parallel to the front of an enemy's position, supposing in the latter case the theatre of operations to be in a friendly country, and the wind, of course, to be favourable. This question will be more fully dealt with in the next chapter.

General observations.

The flatter a country is the more favourable it is for the useful employment of captive balloons; in an undulating country an observer on a hill-top can often see as far as an observer in the car of a balloon; but in a flat country, especially if partially wooded, the advantages to be obtained by the use of a captive balloon are unquestionable.

In this respect the Soudan was an almost ideal country for military ballooning, but unfortunately want of transport and gas seriously hampered ballooning operations in the campaign of 1885.

Moral effect of using balloons.

The moral effect produced by the employment of balloons should not be lost sight of in estimating their value for war purposes.

A captive balloon with nothing but sand-bags or a dummy in the car might paralyze an enemy's initiative, and especially might deter him from attempting any wide turning movement, by causing him to believe that all his movements were being watched.

Chapter VII.

FREE-RUN RECONNAISSANCE.*

The importance of utilising free balloons for reconnaissance work in war has been perhaps somewhat overlooked, chiefly owing to the fact that, except under the unusual circumstances of the siege of a large capital, such as of Paris in 1870-71, very few attempts have been made to use free balloons in warfare at all.

Since, however, it is an accepted axiom that one of the most vitally important factors in the successful conduct of a campaign is the obtaining of reliable information, on which the plans of the General Officer Commanding are based, it is obvious that no *possible* means should be left untried which might, under favourable conditions, prove of inestimable advantage to the nation which was far seeing enough to have worked out during peace time everything connected with the employment of free balloons, and thus be in a position to obtain the fullest advantage from their use in war.

The writer would wish it to be known that the suggestions made and plans of action here laid down are of a purely tentative nature, and are put forward rather as a means of directing the attention of military men to the possibilities of free-run work for reconnaissance purposes than as a definite scheme of how such work should be carried out.

Owing to the fact that free balloons would be most generally used in a civilised, and consequently a well mapped, region, it is clear that their sphere of useful action in respect to the reconnaissance of the topographical features of a country would be extremely limited. Cases might arise in which it might be required to obtain further information of the topographical features of a piece of country which had not been properly or adequately surveyed for military purposes, but since the outcome of all free-run work must naturally be a descent in a friendly country, and a report on the information thus gained, it is evident that a free run over a little known and hostile belt of country with a view to a landing in a friendly zone, would rarely be practicable. {Reconnaissance of a country.}

No such argument would, however, apply to a free run across a hostile belt of country in temporary occupation of the enemy, with the object of observing and reporting on his general movements, lines of advance, positions, &c. Taking into consideration the facilities afforded by railways for transporting balloons and their equipment to any required point in the theatre of operations, and the still greater facilities afforded by the {Reconnaissance of the enemy.}

* Contributed by Major Willoughby Verner, Rifle Brigade.

CHAP. VII.   MANUAL OF MILITARY BALLOONING.

telegraph for transmitting any information thus obtained, it would appear that for this branch of reconnaissance work free balloons might often be employed with advantage.

Since every military operation is necessarily regulated by its own peculiar topographical and strategical problems, it would be clearly undesirable to attempt to generalise here on the best means of employing free balloons in warfare. Rather would it seem more suitable to select some definite example, and to demonstrate how, in such a case, free balloons might conceivably be of great value to one side, leaving it to those who have to work out other problems on the same subject to modify or adapt the procedure here described to the particular case before them.

The example taken is that of a landing on our southern coasts within a zone of, say, from Newhaven to Shoreham, and the subsequent advance of the invaders on the capital. In such a contingency, balloons with the necessary amount of tubes, &c., might be taken by rail to various suitable points on the *windward* side of the enemy's line of advance, such points to be carefully regulated with reference to the direction of the wind on the days in question as ascertained from the weather charts, &c. After making a free run across the belt of country occupied by the enemy, and ascertaining all that could be observed as to the roads taken by the various columns, &c., a descent would be made in a part of the country beyond the reach of the invaders, and the information thus obtained telegraphed to headquarters. Such an operation as the preceding would be perfectly practicable in the counties of Kent, Sussex, and Hants so long as the wind were not northerly or southerly. It would be easy to amplify this subject, but enough has been said to generally indicate the methods of scouting with free balloons. The exact plan to be followed would, of course, vary with each particular case.

Keeping a "dead reckoning" in a balloon.

As regards topographical work in a free balloon, whether it be the careful keeping of a route followed with or without the aid of existing maps, the principle to bear in mind is that all calculations must be made with reference to some *back station*, and not by attempting to observe anything which may *appear* to lie ahead. Thus, in taking the compass bearing of the line of advance of a balloon, it is essential that some well-defined object, such as a building, tall tree, angle of a wood, &c., should be carefully noted at the time the balloon is exactly over it, and its magnetic bearing taken with a compass a few minutes later. This gives the "back bearing," and the true course of the balloon is, of course, the reading thus taken, $\pm 180°$, or 16 points of the compass. For example, a back bearing of 315° would give a course of 135°, or S.E. For rapid observations in a balloon, it is useful to have a specially made compass-card, so that the course may be read at sight from it by observing the back bearing, thus obviating the preceding calculation.

| FREE-RUN RECONNAISSANCE. | CHAP. VII. |

In keeping a record of a free run with the aid of a map in a civilised country, the chief land-marks which form successive points of departure, so to speak, are the railways, next to these come the rivers, and lastly, the main roads. Naturally, with a 1-inch Ordnance map, every yard of the country traversed over can be identified, and the position on it noted. But for ordinary purposes this is far too large a scale map to employ, and the reduced Ordnance, at a scale of 4 miles to 1 inch, will be found to supply every reasonable want. With a map of this description, the exact point where a balloon crosses a railway, river or stream, canal, or cross roads, can be marked from time to time, and the course thus taken drawn in with almost absolute accuracy. If, at the same time, the altitudes at certain points be also jotted down, the relative courses taken by the balloon at varying altitudes can be noted, and from thence a good general idea obtained as to the direction of the upper and lower currents travelled in. For example, during a free run of close on 100 miles, over 50 were made at an altitude of about 2,000 feet on a course of about N.W. Rain coming on, the balloon was taken 1,000 feet higher, so as to avoid the clouds as much as possible, and here entered a southerly current, in which it ran for 30 miles; subsequently, on descending below the clouds, the former south-easterly current was picked up and carried on until the descent was made. *Dead reckoning with aid of maps.*

In this run, no map of the country beyond one of the vicinity of Aldershot was taken, but the course followed was plotted on the map in "Bradshaw's Railway Guide" (scale, 20 miles to 1 inch), by observing the different lines of railway crossed, and now and again getting information by various methods, such as noting the position of a single branch line, a terminus, junction, or river, or other unmistakable point.

In keeping a dead reckoning it is very necessary to ascertain the approximate rate at which the balloon is travelling; in the foregoing example this was found to be about 30 miles an hour, and, since Bradshaw's map is on a scale of 20 miles to 1 inch, this gave about 20 minutes run to every ½ inch on the map. Knowing this, it was not difficult to identify any line of railway passed over by measuring the number of inches from the last well-identified "point of departure," and noting the number of minutes which had since elapsed.

When, for purposes of training the eye to rapid observation of a country from a balloon, or for other reasons, a sketch of the country passed over is required, the following system has been found to give admirable results. *Sketching from a free balloon.*

The first requisite for accurate work to scale is a *true* wind, and, for the best results, one not less than 10, or more than 30, miles an hour is most convenient, in fact, the ordinary wind experienced in fair weather.

The process of drawing in the detail of ground as seen

(M.B.)   F

below the balloon is extremely simple, in so far that the only necessity is to *set* the sketch by some back station, and then, as the balloon is "plumb" above any roads, woods, &c., to draw in by eye the trend of these with relation to the "back station," or line traversed by the balloon. The detail thus drawn in is carried out as far on either side as the eye can be trusted to estimate. For example, supposing a road be crossed 1 mile after passing a railway, objects on the right and left can be drawn in with very fair accuracy by quickly estimating whether they are a mile, or more, or less (*i.e.*, the same distance, or more, or less, according to circumstances, as the railway appears to be). And the lofty and commanding position of the observer enables such an approximation to be made with considerably more accuracy than some would imagine, since the eye can be used as an extemporised clinometer, so to speak, to estimate the relative angles of depression of the objects as they are drawn in, and thus obtain their distance.

The first essential for the execution of a good sketch in a free balloon is a long strip of paper, ruled with a thick line down the centre, to indicate the course taken by the balloon, and with parallel thinner lines on either side of this line, at 1-inch intervals, for the purpose of drawing detail on either side of the course taken. Besides this, the paper is ruled transversely at 1-inch intervals for the purpose of drawing in the sketch by "scale of time."

For ordinary runs, from 10 to 15 feet of paper is sufficient, and the Service tracing cloth (used on the *unglazed* side) will be found useful for this purpose, since it is far more compact than ordinary paper, and, in consequence, permits of a greater ength being carried.

This long strip of paper is placed on a Cavalry Sketching Board, and wound up in the *reverse* way, usually adopted when sketching on horseback. This is because the sketching must all be done, so to speak, "*in arrear*" and *not* by looking ahead, as is the case on horseback. (In other words, the paper at the commencement must all be rolled on the roller *nearest* the sketcher).

Assuming that the board is 7½ inches wide (and it should not be wider for use on active service), there will be a little over 3½ inches on either side of the thick central line denoting the course of the balloon. If the left outer column be reserved for noting "time," "altitudes by barometer," and "compass bearings," the right can be used for observations of "temperature," whether the balloon is "rising or falling," "expenditure of ballast," &c.; this leaves the six inner divisions clear for the sketch, and assuming the scale to be 1 inch to a mile, a belt of country 6 miles in width can be mapped in. Of course, any important objects beyond the 3 mile limit, such as a range of hills, lakes, rivers, railways, towns, &c., can be shown in the usual way by drawing an arrow in the general direction of the object, and writing, for example, "large town, 8 miles," along it.

For the benefit of those unaccustomed to sketching by "scale of time" (a process, by the way, which is, in most cases, the *only* system possible during active military operations, or for explorers on the march), the general method of working will here be described. The *rate* of the balloon having been approximately obtained as hereafter described (see p. 68); to sketch at a scale of 1 inch to a mile, each division on the paper is assigned a value in minutes representing the time occupied in traversing 1 mile. Thus, for example, if the rate of the wind be 10 miles an hour, each 1 inch division would be given a value of six minutes, since in one hour's run there would be 10 such divisions of 1 inch on the paper, representing 10 miles on the ground at a scale of 1 inch to a mile.

Again, supposing the wind were to be found to be 15 miles an hour, here each 1 inch division on the paper would be marked as four minutes, since in one hour's run, 15 of the 1-inch divisions on paper, representing altogether 60 minutes, would be used, and 15 miles of ground would be traversed.

The rough practical rule for assigning the correct value of minutes to each 1 inch division on the paper, given the rate of the wind per hour, is as follows:—

Divide the number 60 by the rate per hour in miles of the wind and the quotient will give the correct value *in minutes* of 1 inch on paper or 1 mile on the ground.

Example.—Wind ascertained to be 20 miles an hour. What value should each inch division be given? Here $\frac{60}{20}$ miles an hour = 3 minutes.

Obviously, if it be required to sketch on a smaller scale, such as 2 miles to 1 inch, *i.e.*, at half the scale; the value in minutes of the 1 inch intervals on paper should be *doubled*.

Example.—To sketch at a rate of 20 miles an hour at a scale of 2 miles to 1 inch, each 1 inch division on paper would represent 2 miles, and 10 inches would represent 20 miles, and the value in time assigned to each such division would be six minutes.

Conversely, to sketch on a larger scale, such as 2 inches to a mile, each division of 1 inch would have to be given *half* the value in minutes assigned for a scale of 1 inch to a mile.

Example.—To sketch at 10 miles an hour on a scale of 2 inches to a mile, 20 inches on paper would represent 10 miles on the ground, and each 1 inch division would be marked as three minutes run.

The broad rule to always keep in mind is that in order to sketch at any uniform scale, the *stronger* the wind the *smaller* the value assigned to each 1 inch division, since it is

CHAP. VII.   MANUAL OF MILITARY BALLOONING.

traversed in a shorter time, and, conversely, the *less* strong the wind the *greater* the value assigned to each 1 inch division.

A table of the values to be assigned to each 1 inch division in balloon sketching, at what may be taken as the normal scale, *i.e.*, 1 inch to 1 mile, is here appended in further explanation of this subject.

Scale for Time Sketching.

Rate of wind per hour in miles.	Divisions on paper.		Scale to 1 mile in inches.
	1 inch.	1 hour.	
5	12 min.	5 in.	1
7¼	8 ,,	7½ ,,	1
10	6 ,,	10 ,,	1
12	5 ,,	12 ,,	1
15	4 ,,	15 ,,	1
20	3 ,,	20 ,,	1
24	2½ ,,	24 ,,	1
30	2 ,,	30 ,,	1
40	1½ ,,	40 ,,	1

To ascertain the rate of the wind for sketching.

In order to form some idea of the rate of the wind, before starting, send up several pilots and estimate by eye the rate at which they appear to be travelling when at the required altitude.

This is, of course, only an approximation, but a useful one, and sometimes with skilled observers, is very fairly accurate.

Say that the rate is estimated to be 12 miles an hour, here the value of each 1 inch division, sketching at 1 inch to a mile, is $\frac{60}{12} = 5$ minutes. Now mark about the third or fourth line from the end of the roller 0, and the next few lines, 5, 10, 15, 20, and for convenience of future reference for sketching, set your watch to an even hour such as 10 A.M. as soon as the balloon is released (your aeronaut must tend to the balloon of course) noting on the sketch at once "time 43 minutes fast," or whatever it may be (see Plate II). The rate of a balloon at the start is of course very irregular, ground currents and eddies are certain to affect it, and it is not until you have got to your first position that any very accurate work can be hoped for.

A good plan when starting from a locality of which a map is available, is to mark off a certain number of miles on the map in the direction taken by the pilots and as soon as the balloon is well up and away, note the time taken in running a measured mile or two, or even four. If this is found to give the scale as estimated before starting, nothing remains to be done beyond continuing the sketch with the 1 inch intervals

Plate II. *To face p. 68*

Portion of Sketch on Cavalry Board, commenced on a scale with wind estimated to be 12 Miles an hour. On reaching B, the rate of wind is found to be 15 Miles an hour and Scale is amended as shewn.

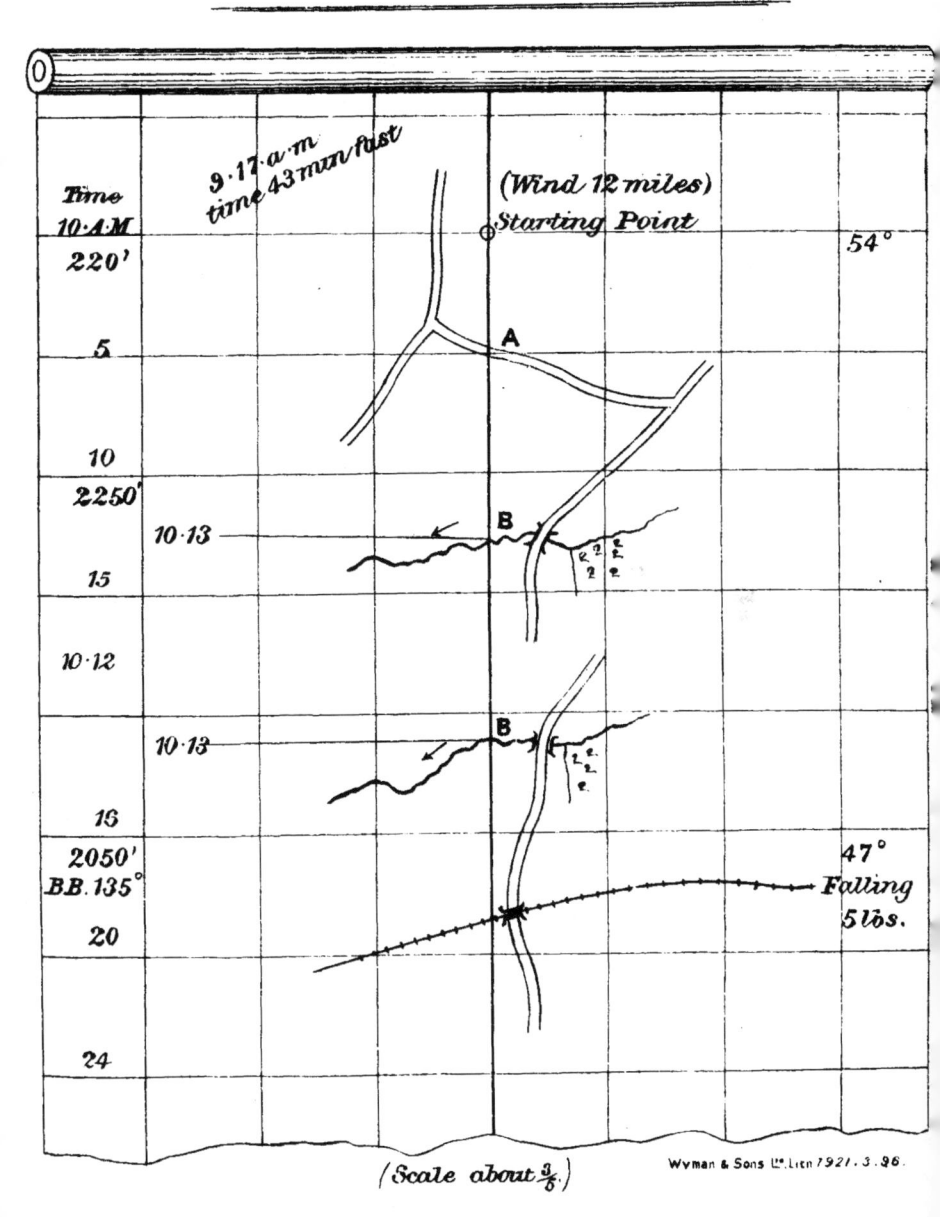

FREE-RUN RECONNAISSANCE.   CHAP. VII.

marked as above described. But suppose, as will very commonly occur, that the first two miles thus measured, say between a road "A" (crossed at 10—5 A.M.), and a stream "B," show that the rate of the wind has been wrongly estimated, and that the latter is crossed at 10—13 A.M., in place of at 10—15 A.M. as estimated; in other words, suppose the rate is found to be 2 miles in 8 minutes, and not 2 miles in 10 minutes. It now becomes necessary to *quickly* grasp the change of scale and act accordingly. It is clear that the rate is quicker than was estimated and further that it is quicker in the proportion of 4 to 5, and therefore that the rate of the balloon is 15 miles an hour in lieu of 12 miles. Hence the true value of each 1 inch division on the sketch is $\frac{60}{15} = 4$ minutes and *not* 5 minutes. To change the scale, write opposite the point where the measurement was taken (the stream) the exact minute noted and draw in the stream B by eye on the old scale ⅗ of the way between 10—10 and 10—15, writing 10—13 in the margin. Now omit a space of the 1 inch divisions and deleting the 20 mark the next one 10—12, and draw the stream B ¼ of the way between 10—12 and 10—16 as shown and continue the sketch, marking the successive intervals as *four minutes* intervals, viz., 10.16, 10.20, 10.24, or rather: 16, 20, 24, &c.

Subsequently, the first few miles on the incorrect scale can be reduced to the correct scale as ascertained, and the sketch completed.

In thus "lifting" the scale over a difficulty, always select some multiple which will carry the divisions on to the hour thus 10—12 was all right, but 10—13, or 10—14 would have brought the sketch up to 11—1 or 11—2. This is not very important, but is worthy of consideration since it eliminates a very common source of error.

This changing of scale requires, of course, some promptness, but it is no very difficult matter, and the results obtained by following the method indicated have been proved to be excellent.

Thus in one run the wind was estimated to be 18 miles an hour before the start, and the inch divisions were in consequence marked 3 minutes, giving a scale of 1·1 inch per mile. After running a few miles it was found that the rate was 30 miles an hour, and the inch divisions were consequently valued at 2 minutes, and the sketch thus carried out for over 50 miles had an error of something under 1 mile.

If no map of the country be available, and in the absence of any means of ascertaining the rate of a free balloon, the process of sketching must be very uncertain, and the only method of ascertaining the scale would be to find the total distance run from a small scale map, and pentagraph in the reconnaissance sketch to the correct scale, a well known process with explorers' route-sketches. Experiments are now in hand at devising some simple instrument to enable the occupants of a balloon to approximately calculate the rate they are travelling at, and when this has been arrived at there will no longer be any difficulty in making a fairly accurate sketch of the country passed over in a balloon without having recourse to the aid of a measured distance at the start.

A word as to the varying rates of speed. With a fair sketching breeze, a balloon that once attains to its elevation, and is in equilibrium would appear to travel at a very uniform rate *so long as this altitude is maintained;* of course by rising or falling a new current may be entered which not only differs in force but in direction from the one hitherto travelled in, and in consequence upsets all calculations. It is always well to note on a sketch any imagined change of rate of travelling as a guide to subsequent correction of the work carried out.

Lastly, as to keeping a record of the bearings by compass of the line taken by the balloon. For general purposes it is a good plan to adjust the index line on the compass of the cavalry sketching board, so that when the board is set the index coincides with the magnetic needle when at rest. This is best done a few minutes after starting, the point whence the ascent was made being taken as the objective.

But it is also necessary to take with an ordinary magnetic compass the "back bearing" of conspicuous objects passed over from time to time, and to note the same in one of the side columns thus: "B.B. clump 235°."

The magnetic needle in the cavalry board will be a good general guide as to whether the original course is being followed, since, when this is the case and the ruler is aligned on the back station, the index bar will also coincide with the needle. Should it not do so, recourse must be had to the compass and the bearing taken. The board is best set on the back bearing by means of a straight-edged ruler *attached* to it, else it may be dropped. A simple and effective means is to lash or screw a straight-edged ruler to the edge of the board so as to project for six inches or so beyond the roller.

When a large car is in use, a wooden seat, suspended from the ring attached to the hoop, is a great convenience, as it enables the reconnoitrer seated on it to keep his eye on any distant point should the balloon continue to revolve, as it often does.

This seat is also a great assistance when using field glasses. The observer seated on it, by means of his feet pressed against the sides of the basket, turning himself round the reverse way to that in which the balloon is revolving, so as to maintain his relative position with the object he is observing.

Chapter VIII.

PHOTOGRAPHY FROM BALLOONS.

Photographs from balloons would be useful in the following cases :— *Aërial photography.*

1. In siege warfare—

    To show the lines and positions of the enemy up to a distance of about 3 miles.
2. To illustrate in peace time the narrative written after manœuvring operations by showing the formation of troops at various times, within a radius of 1 or 2 miles in the open round the balloon, and the general lie of the country.

To obtain good results, it is essential that the exposures should be made under favourable conditions, viz :— *Conditions.*

1. A clear atmosphere.
2. A bright light.
3. Velocity of wind not exceeding 10 miles per hour.

When the above are fulfilled, good photographs should be obtained with a considerable degree of certainty. The usual altitude of the captive balloon would be from 800 to 1,200 feet.

It is difficult to obtain good results when the above conditions are not present.

In hazy or misty weather, little will be seen of the country on the negative for more than a few hundred yards.

In dull weather the image is flat and lacks contrast.

In windy weather—

When the wind velocity exceeds 10 miles an hour, many of the exposures will be somewhat blurred through the movement of the balloon.

When the wind velocity exceeds 15 miles per hour, most of the plates will be spoilt.

With the wind velocity over 20 miles, it is almost useless to attempt to photograph, as the image will be blurred even with an exposure of $\frac{1}{100}$th of a second.

Advantage should be taken of the lulls to obtain an exposure.

In good photographs, troops would be shown up to a distance of a mile in any formation in the open, and formed bodies would be distinguished up to nearly 2 miles. Owing to the manner in which photography always seems to magnify *Capabilities of photography.*

CHAP. VIII.   MANUAL OF MILITARY BALLOONING.

             distance, the figures appear very small in the photograph, and in no case is the camera likely to discover troops not previously seen. The general lie of the country is shown up to a distance of 3 or 4 miles; beyond this, the size of even large objects becomes very microscopic.

Taking the photo.   When it is desired to take a photograph from a captive balloon, the photographer may ascend either alone or with an aeronaut; if alone, he must pay attention to the weather and the state of the balloon itself: if accompanied by an aeronaut, the latter looks after the balloon and sits in the netting. One man with a large camera is quite enough to fill up the car, and the photographer must not be more cramped for space than is necessary.

    The balloon should have a good lift—from 50 lbs. to 100 lbs.—in order that it may be as steady as possible. The grapnel, grapnel rope, and one bag of ballast must always be taken up in the car. The photographs should be taken and the balloon hauled down as speedily as possible. The camera should be firmly tied to the captive rope attachment by a strong cord about 6 feet long, so that this shall not interfere with the moving of the camera.

    After the exposed plates are sent down, plenty of time should be allowed for preparing the negative and the print. One to two hours for developing and drying the negative, and one hour, at least, for the print—toned, slightly washed, and partially dried; the time required for printing, however, depends entirely on the light, and on a dull day, this process may take three or four hours.

Length of exposure.   According to the published tables, the exposure with stop F. 11 about midday in June for ordinary plates for an open landscape, is about $\frac{1}{15}$th second. As, however, it very seldom happens that the camera is at rest, this exposure gives blurred images. An exposure of $\frac{1}{30}$th second has been found to be the most that can ever satisfactorily be used: $\frac{1}{50}$th to $\frac{1}{60}$th of a second is that generally used. On bright summer days, $\frac{1}{100}$th second gives quite a long enough exposure.

    Any lengthening of the exposure, even on dull, calm days, does more harm than good.

Plates.   Ilford Ordinary or Cadett Ordinary are equally suitable on a bright day.

    Ilford Rapid are good on a dull day. Ilford Special Rapid, and Cadett Lightning are decidedly inferior, the additional sensitiveness only causing fog without giving more detail.

    In hot weather, Cadett plates have a much greater tendency to frill than the Ilford plates.

Developer.   Several developers seem to give equally good results.

    A good pyro-ammonia developer is given as under. Each ounce of the liquid containing:—

Pyrogallic acid	..	..	..	3 grains.
Sulphite of soda	..	..	..	6 ,,
Citric acid..	..	..	..	⅕th grain.
Potassium bromide		..	..	1 ,,
Concentrated ammonia	..		..	2½ minims.

A hydroquinon, or pyro-soda, developer of the usual formula is also very satisfactory.

The developing must be slowly and carefully done. It is best to use a weak developer at first, then, as the image appears, to substitute a stronger one, with the ingredients regulated as the plate seems to require. Care must be taken to get plenty of contrast without fog, for the class of negative obtained depends as much, or more, on the developing than on anything else. Open landscape photographs require more attention in developing than do most others. A good, safe light to develop by should be a *sine quâ non*. Development.

The alum bath may be generally omitted to save time, except in very hot weather. Alum bath.

Fix as usual, with hyposulphite of soda. Fixing.

Eastman's "Solio" paper, and Ilford's "Printing-out" paper, are equally good. The prints may be finished off with a combined toning and fixing bath. For working at night, Eastman's "Nikko" paper is good, but does not give quite such full detail as the Solio or P.O.P. Bromide paper is inferior to the above. The platinotype and the ferro-prussiate processes are both practically useless. Printing.

The camera now in use consists of a 1/1 plate, 8½ inches by 6½ inches, fitted with a lens of 16-inches focal length, a Thornton-Pickard extra rapid shutter, working from $\frac{1}{20}$th to $\frac{1}{100}$th of a second, a focussing rack and pinion, and horizontal and vertical swingbacks. The camera complete folds up to a size of 8" × 9" × 10", and with three double dark slides or changing box, weighs 13¼ lbs. The camera.

The lens is a 16-inch rapid rectilinear, by Watson and Sons, fitted with Iris diaphragm and extra rapid Thornton-Pickard shutter, which covers the plate fairly well with stop F. 8, but better results are obtained with F. 11·3. Lens.

A changing box, containing 12 plates or films has been used, but it does not always work smoothly in the balloon. A number of double dark slides, though heavier, are preferable for work in the field; these are safer in use as regards "white light" fogging the plates. Changing box.

When a balloon is free, the unceasing swaying motion of the captive balloon is replaced by a very much less objectionable rotating movement, which is very gentle compared with the oscillations of a captive balloon on a rough day; photography is consequently easier, and the velocity of the wind is not of such importance. Taking $\frac{1}{60}$th second as the usual Photography from free balloons.

CHAP. VIII.   MANUAL OF MILITARY BALLOONING.

exposure, and focal length of lens 16 inches, the balloon would have to travel at a velocity of 60 miles an hour to give a blurr on the plate of $\frac{1}{100}$th of an inch for objects not more than half a mile off; for more distant objects the blurr would be even less. Any blurr there may be is usually caused by the motion of rotation or by not holding the camera steady. Good photographs can therefore be taken with considerable certainty at whatever rate the wind may be blowing, provided the light be bright and atmosphere clear.

Objects of the free run.   A free run would usually be used to ascertain the position and dispositions of the enemy, and endeavours would be made to start the balloon so that it would drift over or near the enemy's supposed position, photographs of which would be taken to accompany the report. In making the exposures, it should be remembered that the angle of view, viz., 28°, includes half a mile of country at a distance of one mile from the camera. The successive exposures must, therefore, be made at points not more than half a mile apart, to obtain a continuous picture. With wind at 15 miles an hour, it would thus be necessary to expose the plates at intervals of from $1\frac{1}{2}$ to 2 min.

Altitude of the balloon.   This can easily be done. The balloon should travel at an altitude of about 3,000 ft., so that each photo shall take in a good extent of country in plan, and the balloon will be out of range of the enemy's bullets. If it is only desired to have photographs of the country a mile distant or more, the intervals between exposures will be longer, but the country is not well shown beyond three miles.

Thus if a free balloon be allowed to drift past a besieged work or position not more than a mile or two off, a photographer in the car would, under favourable photographic conditions be able to take a series of views to illustrate the enemy's works or positions; the distant objects of which would fit together sufficiently well to give a fairly good continuous picture.

APPENDIX A.

MISCELLANEOUS TABLES, ETC., CHIEFLY COMPILED FROM COLONEL COOKE'S R.E. AIDE MEMOIRE AND TRAUTWINE'S CIVIL ENGINEER'S POCKET-BOOK.

TABLE I.
Wind.

Pressure varies as square of velocity.
V = velocity of wind in miles per hour.
v = ,, ,, feet per second.
P = pressure in lbs. per square foot.
a = angle of incidence of direction of wind with plane of surface.
P = 0·00492V² = 0·0023v² = 0·0023v² sin a.
V = √200P.

Miles per hour.	Feet per minute.	Feet per second.	Force in lbs. per square foot.	Description.
1	88	1·47	0·005	Hardly perceptible.
2	176	2·93	0·020	} Just perceptible.
3	264	4·4	0·045	
4	352	5·87	0·080	
5	440	7·33	0·123	
10	880	14·67	0·492	
15	1320	22·0	1·107	Light breeze.
20	1760	29·3	1·968	Gentle breeze.
25	2200	36·6	3·075	} Moderate breeze.
30	2640	44·0	4·428	
35	3080	51·3	6·027	Fresh breeze.
40	3520	58·6	7·872	} Stormy breeze.
45	3960	66·0	9·963	
50	4400	73·3	12·300	Moderate gale.
60	5280	88·0	17·712	Strong gale.
70	6160	102·7	24·108	Storm.
80	7040	117·3	31·488	} Hurricane.
100	8800	146·6	49·200	

Theoretically a solid sphere presents only ⅔ of the resistance to the air opposed by its generating circle; but practically in high winds, on account of the flattening of the balloon, the resistance offered will approximate to that of a plane circle of the same radius as the balloon. Up to 5 miles per hour, the theoretical fraction $\frac{4}{9}$ may be assumed correct. From 5 to 10 miles per hour it may be taken at $\frac{4}{7}$ of the pressure on the plane surface; from 10 to 20 miles per hour at $\frac{4}{7}$; from 20 to 30 miles per hour at $\frac{4}{9}$; and above 30 miles per hour the pressure may be taken as equal to that offered by the plane surface of its generating circle.

TABLE II.

The Atmosphere.

Weight of atmosphere = 14·706 lbs. per square inch.
,, ,, = 29·92 inches of mercury.
,, ,, = 33·7 feet of water.

The barometer falls about $\frac{1}{2}$ inch for every increase of altitude of 500 feet, with a mean temperature of 50° Fahr.

Composition of atmosphere: By volume, 20·8 oxygen, 79·2 nitrogen; by weight, 23 oxygen, 77 nitrogen. It also contains a little ammoniacal gas, and from 3 to 6 parts in 10,000 of its volume of carbonic acid.

TABLE III.

Diurnal Barometric Wave.

Due to the expansion of the column of air by heat of sun and overflowing at top of atmosphere.

The hour of the wave varies with the season from—

January......	Maximum	9 P.M.	and	10 A.M.
to June......	,,	11	,,	9 ,,
January......	Minimum	3	,,	5 ,,
to June......	,,	5	,,	3 ,,

In the tropics it is most apparent, and there is but little variation of period, the average being—

Maximum 10 P.M. and 9.30 A.M.
Minimum 4 ,, 3.30 ,,

The intensity of the wave varies with the latitude.
At the level of the sea the average intensity in inches is—

Latitude N.	0°	5°	18°	24°	30°	35°	40°	45°	52°	57°	62°
Intensity ..	0·09	0·089	0·08	0·07	0·06	0·05	0·04	0·03	0·018	0·009	0

The intensity decreases with the elevation above the level of the sea.

APPENDIX A.

Table IV.

Correction of Barometer for Capillarity. (To be added.)

	in.	in.	in.	in.	in.	in.	in.	in.	in.
Diameter of tube	0·6	0·55	0·5	0·45	0·4	0·35	0·3	0·25	0·2	0·1
Correction unboiled	0·004	0·005	0·007	0·01	0·014	0·02	0·025	0·04	0·059	0·386
,, boiled	0·002	0·003	0·004	0·005	0·007	0·01	0·014	0·02	0·029	0·044

Table V.

Clouds, Varieties of.

1. *Cirrus*, like a feather. 2. *Cirro-cumulus.* 3. *Cirro-stratus.* 4. *Cumulus*, in conical round clusters. 5. *Cumulo-stratus.* 6. *Nimbus*, a rain cloud. 7. *Stratus*, in parallel layers.

Table VI.

Heights from Boiling Point of Water.

Water, at the ordinary pressure of the atmosphere, boils at 212°, but when subjected to higher pressure, as at the bottom of a mine, the boiling point is higher, and when exposed to less atmospheric pressure, as at the top of a mountain, the boiling point is lower. Tables, &c., from Negretti and Zambra.

TABLE I.

Boiling point.	Height in feet.					Boiling point.	Height in feet.				
°	0°·0	0°·2	0°·4	0°·6	0°·8	°	0°·0	0°·2	0°·4	0°·6	0°·8
	ft.	ft.	ft.	ft.	ft.		ft.	ft.	ft.	ft.	ft.
180	17692	17576	17460	17344	17228	197	8087	7977	7867	7758	7648
181	17111	16995	16879	16763	16647	198	7539	7430	7321	7212	7103
182	16532	16416	16300	16185	16070	199	6993	6884	6775	6667	6559
183	15954	15839	15724	15609	15494	200	6450	6341	6233	6125	6016
184	15379	15264	15149	15035	14921	201	5908	5800	5692	5584	5476
185	14806	14692	14578	14464	14350	202	5368	5260	5152	5045	4938
186	14235	14121	14007	13893	13780	203	4830	4722	4615	4508	4401
187	13666	13552	13439	13326	13213	204	4294	4187	4080	3973	3866
188	13100	12987	12874	12761	12648	205	3760	3653	3546	3440	3334
189	12535	12422	12309	12196	12084	206	3227	3121	3015	2909	2803
190	11972	11859	11747	11635	11523	207	2697	2591	2485	2379	2274
191	11411	11299	11187	11075	10964	208	2168	2062	1957	1851	1746
192	10852	10740	10629	10517	10406	209	1640	1535	1430	1325	1220
193	10295	10184	10073	9962	9851	210	1115	1010	905	800	695
194	9740	9629	9518	9408	9297	211	591	486	382	277	173
195	9187	9077	8967	8857	8747	212	69	35	139	243	347
196	8636	8526	8416	8306	8196	213	451	555	659	762	866

APPENDIX A.

Table II.

Mean temp.	Factor.	Mean temp.	Factor.	Mean temp.	Factor.	Mean temp.	Factor.	Mean temp.	Factor.
20	0·973	33	1·002	46	1·031	59	1·060	72	1·089
21	0·976	34	1·004	47	1·033	60	1·062	73	1·091
22	0·978	35	1·007	48	1·036	61	1·064	74	1·093
23	0·980	36	1·009	49	1·038	62	1·067	75	1·096
24	0·982	37	1·011	50	1·040	63	1·069	76	1·098
25	0·984	38	1·013	51	1·042	64	1·071	77	1·100
26	0·987	39	1·016	52	1·044	65	1·073	78	1·102
27	0·989	40	1·018	53	1·047	66	1·076	79	1·104
28	0·991	41	1·020	54	1·049	67	1·078	80	1·107
29	0·993	42	1·022	55	1·051	68	1·080	81	1·109
30	0·996	43	1·024	56	1·053	69	1·082	82	1·111
31	0·998	44	1·027	57	1·056	70	1·084	83	1·113
32	1·000	45	1·029	58	1·058	71	1·087	84	1·116

Rule.—From Table I take out the heights in feet, corresponding to the boiling point observed at the upper and lower stations respectively. The difference between these two numbers, multiplied by the factor in Table II for the mean temperature of the air, is the difference in height required.

Example :—

At upper station, boiling point = 187°·3; temperature of air = 26°.

At lower station, boiling point = 210°·4; temperature of air = 68°.

Boiling point = 187°·3; height from Table I = 13,495 feet.
Boiling point = 210°·4; height from Table I = 905 feet.

Difference = 12,590

Mean temperature of air = 47°; factor from Table II = 1·033.

Required difference between the two stations = 12590 × 1·033 = 13,005 feet.

Table VII.
Weights and Specific Gravities.
Metals.

—	Specific gravity.	Weight of a cubic foot in lbs. avoir.	—	Specific gravity.	Weight of a cubic foot in lbs. avoir.
Aluminium	2·6	162	Gun-metal	8·784	549
Antimony, cast	6·712	419	Iridium	23·000	1437
Arsenic	5·763	360	Iron, wrought	7·0	474
Bismuth, cast	9·822	614		7·8	486
Brass, cast	7·8	487	,, ,, average	7·75	484
	8·4	524		7·0	437
,, wire	8·544	534	,, cast	7·3	450
Bronze	8·218	513	,, ,, average	7·11	444
Cobalt, cast	7·812	488	,, meteoric	7·965	497
Copper, cast	8·788	549	Lead, cast	11·352	709
,, coin	8·915	557	,, milled	11·4	712
,, wire and sheet	8·878	555	Mercury, common, at 32°...	13·568	843
Gold, coin	17·647	1102			
,, trinket	15·609	981	,, pure	14·000	875
,, pure cast	19·258	1203	,, solid	15·6	977
,, hammered	19·316	1210	Nickel	8·279	517

Liquids.

—	Specific gravity.	Weight	—	Specific gravity.	Weight
Alcohol, pure	0·791	49	Oil, whale	0·923	57
,, proof spirit	0·916	57	Petroleum	0·878	55
Ale, average	1·035	64	Proof spirit	0·916	57
Aqua regia	1·234	77	Sulphuric acid	0·848	128
Bitumen, liquid	0·848	53	Tar	1·015	63
Blood, human	1·045	65	Turpentine, oil of	0·870	54
Boracic acid	1·830	114	Vinegar	1·026	64
Brandy	0·837	52	Water, distilled, at 40° Fahr.	1·000	62·5
Ether, muriatic	0·874	54			
,, sulphuric	0·720	45	,, Sea	1·028	64
Milk	1·030	64	,, Mediterranean	1·029	64
Muriatic acid	1·218	76	,, Dead Sea	1·240	77
Naphtha	0·848	53	Wine, Bordeaux	0·994	62
Nitric acid	1·500	93	,, Burgundy	0·991	61
Nitrous acid	1·452	90	,, Champagne	0·997	62
Oil, castor	0·970	60	,, Moselle	0·916	57
,, linseed	0·940	58	,, Port, red	0·990	61
,, olive	0·915	57			

Gases.

—	Specific gravity.	Weight	—	Specific gravity.	Weight
Ammonia	0·000596	0·0368	Muriatic acid	0·001280	0·0800
Atmospheric air	0·001	0·0625	Nitrogen	0·000972	0·0607
Carbonic acid	0·001524	0·0952	Nitrous oxide	0·001042	0·0651
Chlorine	0·002444	0·1527	Olefiant	0·000982	0·0613
Coal gas used in lighting	0·0005	0·0312	Oxygen	0·001111	0·0694
Cyanogen	0·001805	0·1128	Steam at 212°	0·0006	...
Fluosiline acid	0·003611	0·2256	Sulphurous acid	0·002222	0·1388
Hydriodic acid	0·004300	0·2687	Sulphuretted hydrogen	0·00177	...
Hydrogen	0·000069	0·0043			

The above table gives the weight and specific gravity of air as being slightly less than they should be at the normal tem-

APPENDIX A.

perature and pressure. The figures will be found accurate for a pressure of 23·5 inches and a temperature of 40° Fahr.

The specific gravity of air referred to water is more accurately 0·00123 at the normal pressure and temperature, but for simplicity it is shown in the table as 0·001.

To refer the specific gravity of any gas in the above table to air as unity, the decimal point as given in the table should be moved three places to the right.

The following facts and rough figures can be easily carried in the head :—

    1 cubic foot of water weighs 1,000 ozs.
    1 ,, ,, air   ,,   1 oz.
    1 ,, ,, coal gas ,, $\frac{1}{2}$,,
    1 ,, ,, hydrogen ,, $\frac{1}{10}$,,

or the facts regarding hydrogen, air, and water may be stated thus : Weight of 10,000 cubic feet of hydrogen = weight of 1 cubic foot of water = weight of 1,000 cubic feet of air = 1,000 ozs.

TABLE VIII.

Determination of the Specific Gravity of Gases by Schilling's Method.

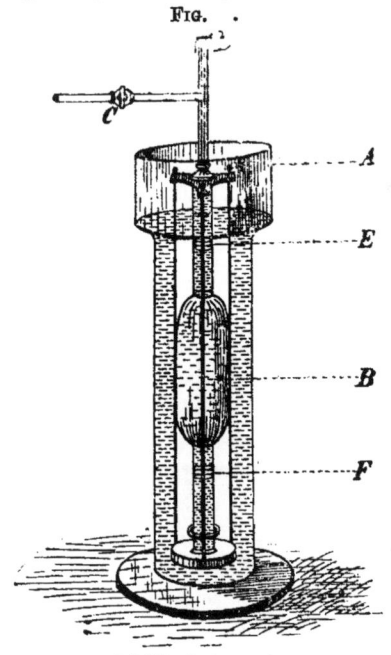

FIG.

Schilling's Apparatus.

(M.B.)

MANUAL OF MILITARY BALLOONING.

Schilling's apparatus consists of a glass vessel, A, containing water, inside which is placed another glass vessel, B, intended to contain the gas to be tested. B is open at the bottom, and is filled with gas through the tap C. The tap C is then closed and the gas is confined under pressure corresponding to the difference of level of the water in A and B.

The tap D is then opened, and the gas escapes through a fine pin-hole in a platinum diaphragm immediately above D.

There are two fixed marks on the inner vessel at E and F. An observation is made of the length of time taken by the water in B to rise from F to E.

The law of diffusion of gases through a small orifice is supposed to be this—that (other things being constant)

$$\text{rate of diffusion} \propto \frac{1}{\sqrt{\text{sp. gr. of gas}}}$$

$$\therefore \text{time of } \quad ,, \quad \propto \sqrt{\text{sp. gr.}}$$

Therefore, if we fill the apparatus with air as described, and allow it to diffuse, and 2nd with hydrogen, and allow it to diffuse, we shall at once obtain the relative densities as follows:—

$$\frac{\text{sp. gr. of H}}{\text{sp. gr. of air}} = \left(\frac{\text{time of diffusion of H}}{\text{time of diffusion of air}}\right)^2$$

$$\text{or sp. gr } H = \left(\frac{t_h}{t_a}\right)^2$$

Table of Specific Gravities for use with Schilling's Apparatus.

$\frac{t_h}{t_a}$	0.	1.	2.	3.	4.	5.	6.	7.	8.	9.
0·28	0·0784	0·0790	0·0795	0·0801	0·0807	0·0812	0·0818	0·0824	0·0829	0·0835
0·29	0·0841	0·0847	0·0853	0·0858	0·0864	0·0870	0·0876	0·0882	0·0888	0·0894
0·30	0·0900	0·0906	0·0912	0·0918	0·0924	0·0930	0·0936	0·0942	0·0949	0·0955
0·31	0·0961	0·0967	0·0973	0·0980	0·0986	0·0992	0·0999	0·1005	0·1011	0·1018
0·32	0·1024	0·1030	0·1037	0·1043	0·1050	0·1056	0·1063	0·1069	0·1076	0·1082
0·33	0·1089	0·1095	0·1102	0·1109	0·1116	0·1122	0·1129	0·1136	0·1142	0·1149
0·34	0·1156	0·1163	0·1170	0·1176	0·1183	0·1190	0·1197	0·1204	0·1211	0·1218
0·35	0·1225	0·1232	0·1239	0·1246	0·1253	0·1260	0·1267	0·1274	0·1282	0·1289
0·36	0·1296	0·1303	0·1310	0·1318	0·1325	0·1332	0·1340	0·1347	0·1354	0·1362
0·37	0·1369	0·1376	0·1384	0·1391	0·1399	0·1406	0·1414	0·1421	0·1429	0·1436
0·38	0·1444	0·1452	0·1459	0·1467	0·1475	0·1482	0·1490	0·1498	0·1505	0·1513

For lifting power of gas see Appendix B, Table I, page 110.

TABLE IX.

Specific Heat.

Is the amount of heat required to bring equal weights or equal volumes of different bodies to the same temperature.

APPENDIX A.

TABLE of Specific Heat of various Substances.

(1) Referred to air as unity.

	For equal volumes.	For equal weights.
Air	1·000	1·000
Hydrogen	0·993	14·353
Carbonic acid	1·350	0·912
Oxygen	1·012	0·914

(2) Referred to water as unity.

Water	1·0000	Platinum	0·0324
Air	0·2669	Tin	0·0562
Bismuth	0·0308	Silver	0·0570
Lead	0·0293	Zinc	0·0955
Gold	0·0324	Copper	0·0951
Iron	0·1137	Sulphur	0·2026
Mercury	0·0333		

If 1 lb. of coal will heat 1 lb. of water to 100°, 0·033 (= $\frac{1}{30}$ lb.) will heat 1 lb. of mercury to 100°.

TABLE X.

Thermometers.

The interval between the temperature of melting ice and that of boiling water is divided into—

    180 parts from 32° to 112° in Fahrenheit.
    100   „   0° „ 100° in Centigrade.
    80   „   0° „ 80° in Reaumur.

$$F = \frac{9}{5} C + 32 = \frac{9}{4} R + 32°.$$

$$R = \frac{4}{9} (F - 32°) = \frac{4}{5} C.$$

$$C = \frac{5}{9} (F - 32°) = \frac{5}{4} R.$$

The zero of Fahrenheit's thermometer corresponds to 14·222, &c., below zero of Reaumur, to 17·777, &c., below zero of Centigrade.

The zero of Wedgwood's pyrometer is 1077° Fahr., and each of his degrees = 130° Fahr.

To test a mercurial thermometer, invert it; the mercury should strike the end with a click, showing there is no air.

Detach a piece of mercury by a jerk and move it about; it

(M.B.)

should read the same length at every part of the tube. Put the thermometer in melting ice, it should read 32° Fahr.; in steam, 212° Fahr.

All thermometers have small errors; where great accuracy is required, they should be compared with standard thermometers for every 10°, the errors noted, and added to or subtracted from the observations.

When parts of the spirit in a spirit thermometer get detached, hold it with the bulb downwards, and give it sharp jerks until the pieces reunite.

Mercurial thermometers may be used between −39° Fahr. and +660° Fahr.

Alcohol may be used in the greatest known cold.

Table XI.

Rules for Weight and Strength of New Government Hemp rope, Wire rope, and Chains.

W = weight in lbs. of rope, in fathoms.
C = circumference of rope, in inches.
S = safe load, in tons.
a = factor of safety = $\frac{3}{20}$. (For new, well-made rope a lower factor of safety may be used, say $\frac{1}{4}$.)
$W = \frac{1}{4} C^2$ for tarred hemp rope (approximate).
$ = C^2$ for iron wire rope (approximate).
$S = \frac{1}{3} aC^2$ for new Government hemp rope.
$ = aC^2$ for iron wire rope.

White rope is about one-seventh lighter, rather stronger, and 40 per cent. less stiff than tarred rope.

Cable-laid wire ropes have less than two-thirds strength of hawser-laid ropes of same circumference, and are about two-thirds weight; they stretch much more, but are more supple.

Steel wire rope is at least twice as strong and a little heavier than iron rope.

The strength of chain varies as the square of the diameter of the iron in the link.

Table of Breaking Weights of Ropes and Chains.

The following table shows the strength and weight of hemp and wire rope, as obtained chiefly by experiment. The strength for hemp being the minimum breaking weights in a series of experiments on new Government rope. These differ somewhat from the results obtained by the formula.

APPENDIX A.

Circumference of rope in inches.	Italian hemp hawser laid (tarred).		Wire ropes, hawser laid.		
	Breaking weight in tons.	Weight per fathom in lbs.	Iron, breaking weight in tons.	Steel,* breaking weight in tons.	Weight per fathom in lbs.
½	0·11	0·15	..	..	..
¾	0·17	0·221	..	..	..
1	0·3	0·3	1·0	..	0·94
1¼	0·89	0·425	1·35	..	1·5
1½	0·94	0·565	2·15	6·25	2·5
2	1·44	0·93	4·0	11·2	3·5
2¼	..	..	5·0	..	4·5
2½	2·16	1·5	6·0	19·5	5·75
2¾	..	..	7·73	..	6·5
3	3·0	2·02	9·2	24·5	7·5
3¼	..	..	10·93	..	8·5
3½	4·2	2·9	12·5	27·5	10·75
4	5·6	3·8	15·75	45·0	13·25
4½	6·75	4·75	21·0	54·5	17·75
5	8·0	6·0	24·8	66·87	21·5
5½	11·0	7·12	30·0	..	26·5
6	14·25	8·50	36·2	100·0	31·5
6½	16·1	10·0	42·75	83·0†	40·6
7	20·0	11·7	48·35	..	42·5
7½	21·75	13·25	55·0	..	46·75
8	25·75	15·0	59·0	..	51·75
8½	28·0	17·0	65·33	..	58·42
9	30·5	19·0	..	..	..
9½	33·75	21·3	..	..	..
10	36·0	23·6	..	..	..
10½	38·9	26·0	..	..	..
11	42·0	28·5	..	..	..
11½	45·1	30·0	..	..	..
12	48·5	34·0	..	..	..

Use a factor of safety of $\frac{3}{20}$, unless for new well-made rope, when ¼ may be used or even ½.

Hemp Ropes.

Italian hemp ropes are stronger than Russian hemp ropes in the ratio of 100 to from 79 to 23·4. On the other hand, the rigidity is in the proportion of Italian 100, Russian 80·4 to 96·5.

Hempen ropes deteriorate after a few months' wear, though apparently good to careful examination with the eye.

With the best ropes not more than six months in use the

* These are the results of Government tests. The wire was from 14 to 19 B.W.G.
† This rope was broken by Kircaldy.

loss was equal to 25 per cent., and extended with other ropes even to 51 per cent., thus showing that it is imperative to use a large margin of safety.

Table of the Qualities of Different Ropes compared with Italian Hemp.

The amount of stretching is that due to the breaking weight. All but the hide ropes are hawser laid.

Nature.	Strength.	Stiffness.	Weight, dry.	Stretching.	Remarks.
Italian hemp	..	1	1	$\frac{1}{7}$ to $\frac{1}{12}$	Become weaker in water. May lose one-third of their strength if soaked for 72 hours.
Baltic „	0·7 to 0·9	0·8 to 0·9	1		
Manilla „	0·9 to 1	0·75	0·88		Keeps its strength in water, but cuts where knotted or bent short; wears in tackle; is difficult to tar.
Flax..	0·9	low	..	$\frac{1}{75}$	..
Coir hair	0·2 to 0·25	„	0·43	$\frac{1}{2}$ to $\frac{1}{3}$	Floats for a time in water; stands exposure to wet well.
Green hide..	0·5	high	1	0·24	When dry does not stretch.
Iron wire	3	„	4	..	..
Steel „	6	„	4	..	..

Large hemp hawsers (8-inch and 9-inch), when new, stretch about $\frac{1}{25}$th, $\frac{1}{13}$th, $\frac{1}{11}$th, and $\frac{1}{9}$th of their length under loads of $\frac{1}{5}$th, $\frac{1}{4}$th, $\frac{1}{3}$rd, and $\frac{1}{2}$ the B.W.; small ropes only stretch from $\frac{1}{3}$rd to $\frac{2}{3}$rds as much. Steel wire hawsers stretch respectively about $\frac{1}{360}$th, $\frac{1}{250}$th, and $\frac{1}{130}$th under a stress of $\frac{1}{4}$th, $\frac{1}{3}$rd, and $\frac{1}{2}$ the B.W.

TABLE XII.

Dip and Distance of Horizon.

d = dip of horizon in seconds.
h = height of observer's eye in feet.
s = distance of horizon in statute miles.
n = „   „   nautical miles.
$d = 57·4 \sqrt{h}$ approximate, varying with temperature.
$h = 0·666 s^2 = ·86 n^2$.

APPENDIX A.

Table of Dip and Distance of Horizon at various Heights.

h.	s.	n.	Dip.	h.	s.	n.	Dip.
feet.	miles.	miles.	′ ″	feet.	miles.	miles.	′ ″
5	2·739	2·411	2 8·35	80	10·959	9·644	8 33·4
10	3·874	3·409	3 1·51	90	11·624	10·229	9 4·54
15	4·745	4·176	3 42·31	100	12·253	10·783	9 34
20	5·480	4·822	4 16·7	150	15·007	13·206	11 43
25	6·126	5·391	4 47	200	17·329	15·249	13 31·76
30	6·711	5·906	5 14·39	300	21·223	18·676	16 34·2
35	7·249	6·379	5 39·58	400	24·507	21·566	19 8
40	7·749	6·819	6 3·03	500	27·399	24·111	21 23·5
45	8·219	7·233	6 25·05	1000	38·749	34·099	30 15·1
50	8·664	7·624	6 45·88	2000	54·799	48·223	42 47
60	9·491	8·352	7 24·62	3000	67·115	59·061	51 23·9
70	10·252	9·021	8 0·24	4000	77·498	68·198	60 30·3

TABLE XIII.

Points of Mariner's Compass and their Angles with the Meridian.

Points.	Angle.	North.		South.	
	° ′ ″				
¼	2 48 45				
½	5 37 30				
¾	8 26 15				
1	11 15 0	N. by E.	N. by W.	S. by E.	S. by W.
1¼	14 3 45				
1½	16 52 30				
1¾	19 41 15				
2	22 30 0	N.N.E.	N.N.W.	S.S.E.	S.S.W.
2¼	25 18 45				
2½	28 7 30				
2¾	30 56 15				
3	33 45 0	N.E. by N.	N.W. by N.	S.E. by S.	S.W. by S.
3¼	36 33 45				
3½	39 22 30				
3¾	42 11 15				
4	45 0 0	N.E.	N.W.	S.E.	S.W.
4¼	47 48 45				
4½	50 37 30				
4¾	53 26 15				
5	56 15 0	N.E. by E.	N.W. by W.	S.E. by E.	S.W. by W.
5¼	59 3 45				
5½	61 52 30				
5¾	64 41 15				
6	67 30 0	E.N.E.	W.N.W.	E.S.E.	W.S.W.
6¼	70 18 45				
6½	73 7 30				
6¾	75 56 15				
7	78 45 0	E. by N.	W. by N.	E. by S.	W. by S.
7¼	81 33 45				
7½	84 22 30				
7¾	87 11 15				
8	90 0 0	E.	W.	E.	W.

Table XIV.
Comparison of Imperial and Metric Systems.
Linear Measure.

	mm.	cm.	Metres.	Kilo-metres.
1 inch	25·4	2·54	0·0254	
1 foot	304·8	30·48	0·3048	
1 yard	914·4	91·44	0·9144	
1 mile	..	..	1609·3	1·609

	Inches.	Feet.	Yards.	Miles.
1 millimetre	0·039	0·0328	0·109	
1 centimetre	0·393	3·281	1·093	
1 metre	39·371	3280·899	1093·63	0·00062
1 kilometre	39370·79			0·621377

Square Measure.

	Sq. mm.	Sq. cm.	Sq. metre.
1 square inch	645·1	6·451	0·00064
1 square foot	92901·0	929·01	0·0929
1 square yard	836112·0	8361·12	0·8361

	Sq. in.	Sq. ft.	Sq. yds.
1 square millimetre	0·00155	0·001076	
1 square centimetre	0·155	10·7641	
1 square metre	1550·03		1·196

APPENDIX A.

Cubic Measure.

	Cub. mm.	Cub. cm.	Cub. metre.	Cub. in.	Cub. ft.	Cub. yds.
1 cubic inch	16387	16·387	0·000016	0·0000610	0·000035	
1 cubic foot	28315300	28315·3	0·0283153	0·06103	35·3166	1·30802
1 cubic yard	764513000	764513·0	0·764513	61027·05		

1 cubic millimetre	0·0000610
1 cubic centimetre	0·06103
1 cubic metre	61027·05

Measures of Capacity.

	Litres.
1 pint	0·56793
1 quart	1·1359
1 gallon	4·5434

1 gramme	1·76 pint
1 kilogramme ..	0·88 quart
	0·22 gallon

$$1 \text{ litre} = \frac{\text{cub. in.}}{61 \cdot 02705} = \frac{\text{cub. ft.}}{0 \cdot 035317}$$

Weights.

	Grammes.	Kilogrammes	Grains.	Lbs.	Cwt.	Ton.
1 grain	0·0648		15·4323	0·0022		
1 ounce	28·349	0·02835				
1 pound	453·59	0·4536				
1 hundredweight ..	50802	50·8			0·01968	
1 ton	1016047·5	1016·047				0·000984

1 gramme	15·4323
1 kilogramme ..	15432·35

1000 kilogrammes = 19 cwts. 2 qrs. 20·6 lbs.

Table XV.

Properties of Circle.—$\pi = 3\cdot 14159$.

Diameter	× 3·14159	= circumference.
Diameter	× 0·886227	= side of an equal square.
Diameter	× 0·7071	= side of inscribed square.
Diameter		= side of a circumscribed square.
Diameter^2	× 0·7854	= area of circle.
Radius^2	× 3·14159	= area of circle.
Radius	× 6·28318	= circumference.
Circumference	× 0·31831	= diameter.
Circumference		= 3·5449 $\sqrt{\text{area of circle}}$.
Diameter		= 1·1283 $\sqrt{\text{area of circle}}$.
Area of circle		= $\pi r^2 = \tfrac{1}{4}\pi d^2$.

$\pi = 3\cdot 14159 = \tfrac{22}{7}$ nearly. Log $\pi = 0\cdot 4971499$.

APPENDIX A.

TABLE XVI.
Table of Circles.

Diameters in Units and Tenths.

Diam.	Circum.	Area.	Diam.	Circum.	Area.	Diam.	Circum.	Area.
0·1	0·314159	0·007854	6·3	19·79203	31·17245	12·5	39·26991	122·7185
·2	0·628319	0·031416	·4	20·10619	32·16991	·6	39·58407	124·6898
·3	0·942478	0·070686	·5	20·42035	33·18307	·7	39·89823	126·6769
·4	1·256637	0·125664	·6	20·73451	34·21194	·8	40·21239	128·6796
·5	1·570796	0·196350	·7	21·04867	35·25652	·9	40·52655	130·6981
·6	1·884956	0·282743	·8	21·36283	36·31681	13·0	40·84070	132·7223
·7	2·199115	0·384845	·9	21·67699	37·39281	·1	41·15486	134·7822
·8	2·513274	0·502655	7·0	21·99115	38·48451	·2	41·46902	136·8478
·9	2·827433	0·636173	·1	22·30531	39·59192	·3	41·78318	138·9291
1·0	3·141593	0·785398	·2	22·61947	40·71504	·4	42·09734	141·0261
·1	3·455752	0·950332	·3	22·93363	41·85387	·5	42·41150	143·1388
·2	3·769911	1·13097	·4	23·24779	43·00840	·6	42·72566	145·2672
·3	4·084070	1·32732	·5	23·56194	44·17865	·7	43·03982	147·4114
·4	4·398230	1·53938	·6	23·87610	45·36460	·8	43·35398	149·5712
·5	4·712389	1·76715	·7	24·19026	46·56626	·9	43·66814	151·7468
·6	5·026548	2·01062	·8	24·50442	47·78362	14·0	43·98230	153·9380
·7	5·340708	2·26980	·9	24·81858	49·01670	·1	44·29646	156·1450
·8	5·654867	2·54469	8·0	25·13274	50·26548	·2	44·61062	158·3677
·9	5·969026	2·83529	·1	25·44690	51·52997	·3	44·92477	160·6061
2·0	6·283185	3·14159	·2	25·76106	52·81017	·4	45·23893	162·8602
·1	6·597345	3·46361	·3	26·07522	54·10608	·5	45·55309	165·1300
·2	6·911504	3·80133	·4	26·38938	55·41769	·6	45·86725	167·4155
·3	7·225663	4·15476	·5	26·70354	56·74502	·7	46·18141	169·7167
·4	7·539822	4·52389	·6	27·01770	58·09805	·8	46·49557	172·0336
·5	7·853982	4·90874	·7	27·33186	59·44679	·9	46·80973	174·3662
·6	8·168141	5·30929	·8	27·64602	60·82123	15·0	47·12389	176·7146
·7	8·482300	5·72555	·9	27·96017	62·21139	·1	47·43805	179·0786
·8	8·796459	6·15752	9·0	28·27433	63·61725	·2	47·75221	181·4584
·9	9·110619	6·60520	·1	28·58849	65·03882	·3	48·06637	183·8539
3·0	9·424778	7·06858	·2	28·90265	66·47610	·4	48·38053	186·2650
·1	9·738937	7·54768	·3	29·21681	67·92909	·5	48·69469	188·6919
·2	10·05310	8·04248	·4	29·53097	69·39778	·6	49·00885	191·1345
·3	10·36726	8·55299	·5	29·84513	70·88218	·7	49·32300	193·5928
·4	10·68142	9·07920	·6	30·15929	72·38229	·8	49·63716	196·0668
·5	10·99557	9·62113	·7	30·47345	73·89811	·9	49·95132	198·5565
·6	11·30973	10·17876	·8	30·78761	75·42964	16·0	50·26548	201·0619
·7	11·62389	10·75210	·9	31·10177	76·97687	·1	50·57964	203·5831
·8	11·93805	11·34115	10·0	31·41593	78·53982	·2	50·89380	206·1199
·9	12·25221	11·94591	·1	31·73009	80·11847	·3	51·20796	208·6724
4·0	12·56637	12·56637	·2	32·04425	81·71282	·4	51·52212	211·2407
·1	12·88053	13·20254	·3	32·35840	83·32289	·5	51·83628	213·8246
·2	13·19469	13·85442	·4	32·67256	84·94867	·6	52·15044	216·4243
·3	13·50885	14·52201	·5	32·98672	86·59015	·7	52·46460	219·0397
·4	13·82301	15·20531	·6	33·30088	88·24734	·8	52·77876	221·6708
·5	14·13717	15·90431	·7	33·61504	89·92024	·9	53·09292	224·3176
·6	14·45133	16·61903	·8	33·92920	91·60884	17·0	53·40708	226·9801
·7	14·76549	17·34945	·9	34·24336	93·31316	·1	53·72123	229·6583
·8	15·07964	18·09557	11·0	34·55752	95·03318	·2	54·03539	232·3522
·9	15·39380	18·85741	·1	34·87168	96·76891	·3	54·34955	235·0618
5·0	15·70796	19·63495	·2	35·18584	98·52035	·4	54·66371	237·7871
·1	16·02212	20·42821	·3	35·50000	100·2875	·5	54·99787	240·5280
·2	16·33628	21·23717	·4	35·81416	102·0703	·6	55·29203	243·2849
·3	16·65044	22·06183	·5	36·12832	103·8689	·7	55·60619	246·0574
·4	16·96460	22·90221	·6	36·44247	105·6832	·8	55·92035	248·8456
·5	17·27876	23·75829	·7	36·75663	107·5132	·9	56·23451	251·6494
·6	17·59292	24·63009	·8	37·07079	109·3588	18·0	56·54867	254·4690
·7	17·90708	25·51759	·9	37·38495	111·2202	·1	56·86283	257·3043
·8	18·22124	26·42079	12·0	37·69911	113·0973	·2	57·17699	260·1553
·9	18·53540	27·33971	·1	38·01327	114·9901	·3	57·49115	263·0220
6·0	18·84956	28·27433	·2	38·32743	116·8987	·4	57·80530	265·9044
·1	19·16372	29·22467	·3	38·64159	118·8229	·5	58·11946	268·8025
·2	19·47787	30·19071	·4	38·95575	120·7628	·6	58·43362	271·7163

MANUAL OF MILITARY BALLOONING.

TABLE XVI.—Table of Circles (continued).

Diameters in Units and Tenths.

Diam.	Circum.	Area.	Diam.	Circum.	Area.	Diam.	Circum.	Area.
18·7	58·74778	274·6459	25·5	80·11061	510·7052	32·3	101·4734	819·3980
·8	59·06194	277·5911	·6	80·42477	514·7185	·4	101·7876	824·4796
·9	59·37610	280·5521	·7	80·73893	518·7476	·5	102·1018	829·5768
19·0	59·69026	283·5287	·8	81·05309	522·7924	·6	102·4159	834·6898
·1	60·00442	286·5211	·9	81·36725	526·8529	·7	102·7301	839·8184
·2	60·31858	289·5292	26·0	81·68141	530·9292	·8	103·0442	844·9628
·3	60·63274	292·5530	·1	81·99557	535·0211	·9	103·3584	850·1228
·4	60·94690	295·5925	·2	82·30973	539·1287	33·0	103·6726	855·5986
·5	61·26106	298·6477	·3	82·62389	543·2521	·1	103·9867	860·4901
·6	61·57522	301·7186	·4	82·93805	547·3911	·2	104·3009	865·6973
·7	61·88938	304·8052	·5	83·25221	551·5459	·3	104·6150	870·9202
·8	62·20353	307·9075	·6	83·56636	555·7163	·4	104·9292	876·1588
·9	62·51769	311·0255	·7	83·88052	559·9025	·5	105·2434	881·4131
20·0	62·83185	314·1593	·8	84·19468	564·1044	·6	105·5575	886·6831
·1	63·14601	317·3087	·9	84·50884	568·3220	·7	105·8717	891·9688
·2	63·46017	320·4739	27·0	84·82300	572·5553	·8	106·1858	897·2703
·3	63·77433	323·6547	·1	85·13716	576·8043	·9	106·5000	902·5874
·4	64·08849	326·8513	·2	85·45132	581·0690	34·0	106·8142	907·9203
·5	64·40265	330·0636	·3	85·76548	585·3494	·1	107·1283	913·2688
·6	64·71681	333·2916	·4	86·07964	589·6455	·2	107·4425	918·6331
·7	65·03097	336·5353	·5	86·39380	593·9574	·3	107·7566	924·0131
·8	65·34513	339·7947	·6	86·70796	598·2849	·4	108·0708	929·4088
·9	65·65929	343·0698	·7	87·02212	602·6282	·5	108·3849	934·8202
21·0	65·97345	346·3606	·8	87·33628	606·9871	·6	108·6991	940·2473
·1	66·28760	349·6671	·9	87·65044	611·3618	·7	109·0133	945·6901
·2	66·60176	352·9894	28·0	87·96459	615·7522	·8	109·3274	951·1486
·3	66·91592	356·3273	·1	88·27875	620·1582	·9	109·6416	956·6228
·4	67·23008	359·6809	·2	88·59291	624·5800	35·0	109·9557	962·1128
·5	67·54424	363·0503	·3	88·90707	629·0175	·1	110·2699	967·6184
·6	67·85840	366·4354	·4	89·22123	633·4707	·2	110·5841	973·1397
·7	68·17256	369·8361	·5	89·53539	637·9397	·3	110·8982	978·6768
·8	68·48672	373·2526	·6	89·84955	642·4243	·4	111·2124	984·2296
·9	68·80088	376·6848	·7	90·16371	646·9246	·5	111·5265	989·7980
22·0	69·11504	380·1327	·8	90·47787	651·4407	·6	111·8407	995·3822
·1	69·42920	383·5963	·9	90·79203	655·9724	·7	112·1549	1000·9821
·2	69·74336	387·0756	29·0	91·10619	660·5199	·8	112·4690	1006·5977
·3	70·05752	390·5707	·1	91·42035	665·0830	·9	112·7832	1012·2290
·4	70·37168	394·0814	·2	91·73451	669·6619	36·0	113·0973	1017·8760
·5	70·68583	397·6078	·3	92·04866	674·2565	·1	113·4115	1023·5387
·6	70·99999	401·1500	·4	92·36282	678·8668	·2	113·7257	1029·2172
·7	71·31415	404·7078	·5	92·67698	683·4928	·3	114·0398	1034·9113
·8	71·62831	408·2814	·6	92·99114	688·1345	·4	114·3540	1040·6212
·9	71·94247	411·8707	·7	93·30530	692·7919	·5	114·6681	1046·3467
23·0	72·25663	415·4756	·8	93·61946	697·4650	·6	114·9823	1052·0880
·1	72·57079	419·0963	·9	93·93362	702·1538	·7	115·2965	1057·8449
·2	72·88495	422·7327	30·0	94·24778	706·8583	·8	115·6106	1063·6176
·3	73·19911	426·3848	·1	94·56194	711·5786	·9	115·9248	1069·4060
·4	73·51327	430·0526	·2	94·87610	716·3145	37·0	116·2389	1075·2101
·5	73·82743	433·7361	·3	95·19026	721·0662	·1	116·5531	1081·0299
·6	74·14159	437·4354	·4	95·50442	725·8336	·2	116·8672	1086·8654
·7	74·45575	441·1503	·5	95·81858	730·6166	·3	117·1814	1092·7166
·8	74·76991	444·8809	·6	96·13274	735·4154	·4	117·4956	1098·5835
·9	75·08406	448·6273	·7	96·44689	740·2299	·5	117·8097	1104·4662
24·0	75·39822	452·3893	·8	96·76105	745·0601	·6	118·1239	1110·3645
·1	75·71238	456·1671	·9	97·07521	749·9060	·7	118·4380	1116·2786
·2	76·02654	459·9606	31·0	97·38937	754·7676	·8	118·7522	1122·2083
·3	76·34070	463·7698	·1	97·70353	759·6450	·9	119·0664	1128·1538
·4	76·65486	467·5947	·2	98·01769	764·5380	38·0	119·3805	1134·1149
·5	76·96902	471·4352	·3	98·33185	769·4467	·1	119·6947	1140·0918
·6	77·28318	475·2916	·4	98·64601	774·3712	·2	120·0088	1146·0844
·7	77·59734	479·1636	·5	98·96017	779·3113	·3	120·3230	1152·0927
·8	77·91150	483·0513	·6	99·27433	784·2672	·4	120·6372	1158·1167
·9	78·22566	486·9547	·7	99·58849	789·2388	·5	120·9513	1164·1564
25·0	78·53982	490·8739	·8	99·90265	794·2260	·6	121·2655	1170·2118
·1	78·85398	494·8087	·9	100·2168	799·2290	·7	121·5796	1176·2830
·2	79·16813	498·7592	32·0	100·5310	804·2477	·8	121·8938	1182·3698
·3	79·48229	502·7255	·1	100·8451	809·2821	·9	122·2080	1188·4724
·4	79·79645	506·7075	·2	101·1593	814·3322	39·0	122·5221	1194·5906

APPENDIX A.

Table XVI.—Table of Circles (continued).

Diameters in Units and Tenths.

Diam.	Circum.	Area.	Diam.	Circum.	Area.	Diam.	Circum.	Area.
39·1	122·8363	1200·7246	45·9	144·1091	1654·6847	52·7	165·5619	2181·2785
·2	123·1504	1206·8742	46·0	144·5133	1661·9025	·8	165·8761	2189·5644
·3	123·4646	1213·0396	·1	144·8274	1669·1360	·9	166·1903	2197·8661
·4	123·7788	1219·2207	·2	145·1416	1676·3853	53·0	166·5044	2206·1834
·5	124·0929	1225·4175	·3	145·4557	1683·6502	·1	166·8186	2214·5165
·6	124·4071	1231·6300	·4	145·7699	1690·9308	·2	167·1327	2222·8653
·7	124·7212	1237·8582	·5	146·0841	1698·2272	·3	167·4469	2231·2298
·8	125·0354	1244·1021	·6	146·3982	1705·5392	·4	167·7610	2239·6100
·9	125·3495	1250·3617	·7	146·7124	1712·8670	·5	168·0752	2248·0059
40·0	125·6637	1256·6371	·8	147·0265	1720·2105	·6	168·3894	2256·4175
·1	125·9779	1262·9281	·9	147·3407	1727·5697	·7	168·7035	2264·8448
·2	126·2920	1269·2348	47·0	147·6549	1734·9445	·8	169·0177	2273·2879
·3	126·6062	1275·5573	·1	147·9690	1742·3351	·9	169·3318	2281·7466
·4	126·9203	1281·8955	·2	148·2832	1749·7414	54·0	169·6460	2290·2210
·5	127·2345	1288·2493	·3	148·5973	1757·1635	·1	169·9602	2298·7112
·6	127·5487	1294·6189	·4	148·9115	1764·6012	·2	170·2743	2307·2171
·7	127·8628	1301·0042	·5	149·2257	1772·0546	·3	170·5885	2315·7386
·8	128·1770	1307·4052	·6	149·5398	1779·5237	·4	170·9026	2324·2759
·9	128·4911	1313·8219	·7	149·8540	1787·0086	·5	171·2168	2332·8289
41·0	128·8053	1320·2543	·8	150·1681	1794·5091	·6	171·5310	2341·3976
·1	129·1195	1326·7024	·9	150·4823	1802·0254	·7	171·8451	2349·9820
·2	129·4336	1333·1663	48·0	150·7964	1809·5574	·8	172·1593	2358·5821
·3	129·7478	1339·6458	·1	151·1106	1817·1050	·9	172·4734	2367·1979
·4	130·0619	1346·1410	·2	151·4248	1824·6684	55·0	172·7876	2375·8294
·5	130·3761	1352·6520	·3	151·7389	1832·2475	·1	173·1018	2384·4767
·6	130·6903	1359·1786	·4	152·0531	1839·8423	·2	173·4159	2393·1396
·7	131·0044	1365·7210	·5	152·3672	1847·4528	·3	173·7301	2401·8183
·8	131·3186	1372·2791	·6	152·6814	1855·0790	·4	174·0442	2410·5126
·9	131·6327	1378·8529	·7	152·9956	1862·7210	·5	174·3584	2419·2227
42·0	131·9469	1385·4424	·8	153·3097	1870·3786	·6	174·6726	2427·9485
·1	132·2611	1392·0476	·9	153·6239	1878·0519	·7	174·9867	2436·6899
·2	132·5752	1398·6685	49·0	153·9380	1885·7410	·8	175·3009	2445·4471
·3	132·8894	1405·3051	·1	154·2522	1893·4457	·9	175·6150	2454·2200
·4	133·2035	1411·9574	·2	154·5664	1901·1662	56·0	175·9292	2463·0086
·5	133·5177	1418·6254	·3	154·8805	1908·9024	·1	176·2433	2471·8130
·6	133·8318	1425·3092	·4	155·1947	1916·6543	·2	176·5575	2480·6330
·7	134·1460	1432·0086	·5	155·5088	1924·4218	·3	176·8717	2489·4687
·8	134·4602	1438·7238	·6	155·8230	1932·2051	·4	177·1858	2498·3201
·9	134·7743	1445·4546	·7	156·1372	1940·0041	·5	177·5000	2507·1873
43·0	135·0885	1452·2012	·8	156·4513	1947·8189	·6	177·8141	2516·0701
·1	135·4026	1458·9635	·9	156·7655	1955·6493	·7	178·1283	2524·9687
·2	135·7168	1465·7415	50·0	157·0796	1963·4954	·8	178·4425	2533·8830
·3	136·0310	1472·5352	·1	157·3938	1971·3572	·9	178·7566	2542·8129
·4	136·3451	1479·3446	·2	157·7080	1979·2348	57·0	179·0708	2551·7586
·5	136·6593	1486·1697	·3	158·0221	1987·1280	·1	179·3849	2560·7200
·6	136·9734	1493·0105	·4	158·3363	1995·0370	·2	179·6991	2569·6971
·7	137·2376	1499·8670	·5	158·6504	2002·9617	·3	180·0133	2578·6899
·8	137·6018	1506·7393	·6	158·9646	2010·9020	·4	180·3274	2587·6985
·9	137·9159	1513·6272	·7	159·2787	2018·8581	·5	180·6416	2596·7227
44·0	138·2301	1520·5308	·8	159·5929	2026·8299	·6	180·9557	2605·7626
·1	138·5442	1527·4502	·9	159·9071	2034·8174	·7	181·2699	2614·8183
·2	138·8584	1534·3853	51·0	160·2212	2042·8206	·8	181·5841	2623·8896
·3	139·1726	1541·3360	·1	160·5354	2050·8395	·9	181·8982	2632·9767
·4	139·4867	1548·3025	·2	160·8495	2058·8742	58·0	182·2124	2642·0794
·5	139·8009	1555·2847	·3	161·1637	2066·9245	·1	182·5265	2651·1979
·6	140·1150	1562·2826	·4	161·4779	2074·9905	·2	182·8407	2660·3321
·7	140·4292	1569·2962	·5	161·7920	2083·0723	·3	183·1549	2669·4820
·8	140·7434	1576·3255	·6	162·1062	2091·1697	·4	183·4690	2678·6476
·9	141·0575	1583·3706	·7	162·4203	2099·2829	·5	183·7832	2687·8289
45·0	141·3717	1590·4313	·8	162·7345	2107·4118	·6	184·0973	2697·0259
·1	141·6858	1597·5077	·9	163·0487	2115·5563	·7	184·4115	2706·2386
·2	142·0000	1604·5999	52·0	163·3628	2123·7166	·8	184·7256	2715·4670
·3	142·3141	1611·7077	·1	163·6770	2131·8926	·9	185·0398	2724·7112
·4	142·6283	1618·8313	·2	163·9911	2140·0843	59·0	185·3540	2733·9710
·5	142·9425	1625·9705	·3	164·3053	2148·2917	·1	185·6681	2743·2466
·6	143·2566	1633·1255	·4	164·6195	2156·5149	·2	185·9823	2752·5378
·7	143·5708	1640·2962	·5	164·9336	2164·7537	·3	186·2964	2761·8448
·8	143·8849	1647·4826	·6	165·2478	2173·0082	·4	186·6106	2771·1675

Table XVI.—Table of Circles (continued).

Diameters in Units and Tenths.

Diam.	Circum.	Area.	Diam.	Circum.	Area.	Diam.	Circum.	Area.
59·5	186·9248	2780·5058	66·3	208·2876	3452·3669	73·1	229·6504	4196·8615
·6	187·2389	2789·8599	·4	208·6018	3462·7891	·2	229·9646	4208·3519
·7	187·5531	2799·2297	·5	208·9159	3473·2270	·3	230·2787	4219·8579
·8	187·8672	2808·6152	·6	209·2301	3483·6807	·4	230·5929	4231·3797
·9	188·1814	2818·0165	·7	209·5442	3494·1500	·5	230·9071	4242·9172
60·0	188·4956	2827·4334	·8	209·8584	3504·6351	·6	231·2212	4254·4704
·1	188·8097	2836·8660	·9	210·1725	3515·1359	·7	231·5354	4266·0394
·2	189·1239	2846·3144	67·0	210·4867	3525·6524	·8	231·8495	4277·6240
·3	189·4380	2855·7784	·1	210·8009	3536·1845	·9	232·1637	4289·2243
·4	189·7522	2865·2582	·2	211·1150	3546·7324	74·0	232·4779	4300·8403
·5	190·0664	2874·7536	·3	211·4292	3557·2960	·1	232·7920	4312·4721
·6	190·3805	2884·2648	·4	211·7433	3567·8754	·2	233·1062	4324·1195
·7	190·6947	2893·7917	·5	212·0575	3578·4704	·3	233·4203	4335·7827
·8	191·0088	2903·3343	·6	212·3717	3589·0811	·4	233·7345	4347·4616
·9	191·3230	2912·8926	·7	212·6858	3599·7075	·5	234·0487	4359·1562
61·0	191·6372	2922·4666	·8	213·0000	3610·3497	·6	234·3628	4370·8664
·1	191·9513	2932·0563	·9	213·3141	3621·0075	·7	234·6770	4382·5924
·2	192·2656	2941·6617	68·0	213·6283	3631·6811	·8	234·9911	4394·3341
·3	192·5796	2951·2828	·1	213·9425	3642·3704	·9	235·3053	4406·0916
·4	192·8938	2960·9197	·2	214·2566	3653·0754	75·0	235·6194	4417·8647
·5	193·2079	2970·5722	·3	214·5708	3663·7960	·1	235·9336	4429·6535
·6	193·5221	2980·2405	·4	214·8849	3674·5324	·2	236·2478	4441·4580
·7	193·8363	2989·9244	·5	215·1991	3685·2845	·3	236·5619	4453·2783
·8	194·1504	2999·6241	·6	215·5133	3696·0523	·4	236·8761	4465·1142
·9	194·4646	3009·3395	·7	215·8274	3706·8359	·5	237·1902	4476·9659
62·0	194·7787	3019·0705	·8	216·1416	3717·6351	·6	237·5044	4488·8332
·1	195·0929	3028·8173	·9	216·4557	3728·4500	·7	237·8186	4500·7163
·2	195·4071	3038·5798	69·0	216·7699	3739·2807	·8	238·1327	4512·6151
·3	195·7212	3048·3580	·1	217·0841	3750·1270	·9	238·4469	4524·5296
·4	196·0354	3058·1520	·2	217·3982	3760·9891	76·0	238·7610	4536·4598
·5	196·3495	3067·9616	·3	217·7124	3771·8668	·1	239·0752	4548·4057
·6	196·6637	3077·7869	·4	218·0265	3782·7603	·2	239·3894	4560·3673
·7	196·9779	3087·6279	·5	218·3407	3793·6695	·3	239·7035	4572·3446
·8	197·2920	3097·4847	·6	218·6548	3804·5944	·4	240·0177	4584·3377
·9	197·6062	3107·3571	·7	218·9690	3815·5350	·5	240·3318	4596·3464
63·0	197·9203	3117·2453	·8	219·2832	3826·4913	·6	240·6460	4608·3708
·1	198·2345	3127·1492	·9	219·5973	3837·4633	·7	240·9602	4620·4110
·2	198·5487	3137·0688	70·0	219·9115	3848·4510	·8	241·2743	4632·4669
·3	198·8628	3147·0040	·1	220·2256	3859·4544	·9	241·5885	4644·5384
·4	199·1770	3156·9550	·2	220·5398	3870·4736	77·0	241·9026	4656·6257
·5	199·4911	3166·9217	·3	220·8540	3881·5084	·1	242·2168	4668·7287
·6	199·8053	3176·9042	·4	221·1681	3892·5590	·2	242·5310	4680·8474
·7	200·1195	3186·9023	·5	221·4823	3903·6252	·3	242·8451	4692·9818
·8	200·4336	3196·9161	·6	221·7964	3914·7072	·4	243·1593	4705·1319
·9	200·7478	3206·9456	·7	222·1106	3925·8049	·5	243·4734	4717·2977
64·0	201·0619	3216·9909	·8	222·4248	3936·9182	·6	243·7876	4729·4792
·1	201·3761	3227·0518	·9	222·7389	3948·0473	·7	244·1017	4741·6765
·2	201·6902	3237·1285	71·0	223·0531	3959·1921	·8	244·4159	4753·8894
·3	202·0044	3247·2209	·1	223·3672	3970·3526	·9	244·7301	4766·1181
·4	202·3186	3257·3289	·2	223·6814	3981·5289	78·0	245·0442	4778·3624
·5	202·6327	3267·4527	·3	223·9956	3992·7208	·1	245·3584	4790·6225
·6	202·9469	3277·5922	·4	224·3097	4003·9284	·2	245·6725	4802·8983
·7	203·2610	3287·7474	·5	224·6239	4015·1518	·3	245·9867	4815·1897
·8	203·5752	3297·9183	·6	224·9380	4026·3908	·4	246·3009	4827·4969
·9	203·8894	3308·1049	·7	225·2522	4037·6456	·5	246·6150	4839·8198
65·0	204·2035	3318·3072	·8	225·5664	4048·9160	·6	246·9292	4852·1584
·1	204·5177	3328·5253	·9	225·8805	4060·2022	·7	247·2433	4864·5128
·2	204·8318	3338·7590	72·0	226·1947	4071·5041	·8	247·5575	4876·9828
·3	205·1460	3349·0085	·1	226·5088	4082·8217	·9	247·8717	4889·2685
·4	205·4602	3359·2736	·2	226·8230	4094·1550	79·0	248·1858	4901·6699
·5	205·7743	3669·5545	·3	227·1371	4105·5040	·1	248·5000	4914·0871
·6	206·0885	3379·8510	·4	227·4513	4116·8687	·2	248·8141	4926·5199
·7	206·4026	3390·1633	·5	227·7655	4128·2491	·3	249·1283	4938·9685
·8	206·7168	3400·4913	·6	228·0796	4139·6452	·4	249·4425	4591·4328
·9	207·0310	3410·8350	·7	228·3938	4151·0571	·5	249·7566	4963·9127
66·0	207·3451	3421·1944	·8	228·7079	4162·4846	·6	250·0708	4976·4084
·1	207·6593	3431·5695	·9	229·0221	4173·9379	·7	250·3849	4988·9198
·2	207·9734	3441·9603	73·0	229·3363	4185·3868	·8	250·6991	5001·4469

APPENDIX A.

TABLE XVI.—Table of Circles (continued).
Diameters in Units and Tenths.

Diam.	Circum.	Area.	Diam.	Circum.	Area.	Diam.	Circum.	Area.
79·9	251·0133	5013·9897	86·7	272·3761	5903·7516	93·4	293·4248	6851·4680
80·0	251·3274	5026·5482	·8	272·6902	5917·3783	·5	293·7389	6866·1471
·1	251·6416	5039·1225	·9	273·0044	5931·0206	·6	294·0531	6880·8419
·2	251·9557	5051·7124	87·0	273·3186	5944·6787	·7	294·3672	6895·5524
·3	252·2699	5064·3180	·1	273·6327	5958·3525	·8	294·6814	6910·2786
·4	252·5840	5076·9394	·2	273·9469	5972·0420	·9	294·9956	6925·0205
·5	252·8982	5089·5764	·3	274·2610	5985·7472	94·0	295·3097	6939·7782
·6	253·2124	5102·2292	·4	274·5752	5999·4681	·1	295·6239	6954·5515
·7	253·5265	5114·8977	·5	274·8894	6013·2047	·2	295·9380	6969·3406
·8	253·8407	5127·5819	·6	275·2035	6026·9570	·3	296·2522	6984·1453
·9	254·1548	5140·2818	·7	275·5177	6040·7250	·4	296·5663	6998·9658
81·0	254·4690	5152·9974	·8	275·8318	6054·5088	·5	296·8805	7013·8019
·1	254·7832	5165·7287	·9	276·1460	6068·3082	·6	297·1947	7028·6538
·2	255·0973	5178·4757	88·0	276·4602	6082·1234	·7	297·5088	7043·5214
·3	255·4115	5191·2384	·1	276·7743	6095·9542	·8	297·8230	7058·4047
·4	255·7256	5204·0168	·2	277·0885	6109·8008	·9	298·1371	7073·3037
·5	256·0398	5216·8110	·3	277·4026	6123·6631	95·0	298·4513	7088·2184
·6	256·3540	5229·6208	·4	277·7168	6137·5411	·1	298·7655	7103·1488
·7	256·6681	5242·4463	·5	278·0309	6151·4348	·2	299·0796	7118·0950
·8	256·9823	5255·2876	·6	278·3451	6165·3442	·3	299·3938	7133·0568
·9	257·2964	5268·1446	·7	278·6593	6179·2693	·4	299·7079	7148·0343
82·0	257·6106	5281·0173	·8	278·9734	6193·2101	·5	300·0221	7163·0276
·1	257·9248	5293·9056	·9	279·2876	6207·1666	·6	300·3363	7178·0366
·2	258·2389	5306·8097	89·0	279·6017	6221·1389	·7	300·6504	7193·0612
·3	258·5531	5319·7295	·1	279·9159	6235·1268	·8	300·9646	7208·1016
·4	258·8672	5332·6650	·2	280·2301	6249·1304	·9	301·2787	7223·1577
·5	259·1814	5345·6162	·3	280·5442	6263·1498	96·0	301·5929	7238·2295
·6	259·4956	5358·5832	·4	280·8584	6277·1849	·1	301·9071	7253·3170
·7	259·8097	5371·5658	·5	281·1725	6291·2356	·2	302·2212	7268·4202
·8	260·1239	5384·5641	·6	281·4867	6305·3021	·3	302·5354	7283·5391
·9	260·4380	5397·5782	·7	281·8009	6319·3843	·4	302·8495	7298·6737
83·0	260·7522	5410·6079	·8	282·1150	6333·4822	·5	303·1637	7313·8240
·1	261·0663	5423·6534	·9	282·4292	6347·5958	·6	303·4779	7328·9901
·2	261·3805	5436·7146	90·0	282·7433	6361·7251	·7	303·7920	7344·1718
·3	261·6947	5449·7915	·1	283·0575	6375·8701	·8	304·1062	7359·3693
·4	262·0088	5462·8840	·2	283·3717	6390·0309	·9	304·4203	7374·5824
·5	262·3230	5475·9923	·3	283·6858	6404·2073	97·0	304·7345	7389·8113
·6	262·6371	5489·1163	·4	284·0000	6418·3995	·1	305·0486	7405·0559
·7	262·9513	5502·2561	·5	284·3141	6432·6073	·2	305·3628	7420·3162
·8	263·2655	5515·4115	·6	284·6283	6446·8309	·3	305·6770	7435·5922
·9	263·5796	5528·5826	·7	284·9425	6461·0701	·4	305·9911	7450·8839
84·0	263·8938	5541·7694	·8	285·2566	6475·3251	·5	306·3053	7466·1913
·1	264·2079	5554·9720	·9	285·5708	6489·5958	·6	306·6194	7481·5144
·2	264·5221	5568·1902	91·0	285·8849	6503·8822	·7	306·9336	7496·8532
·3	264·8363	5581·4242	·1	286·1991	6518·1843	·8	307·2478	7512·2078
·4	265·1504	5594·8739	·2	286·5133	6532·5021	·9	307·5619	7527·5780
·5	265·4646	5607·9392	·3	286·8274	6546·8356	98·0	307·8761	7542·9640
·6	265·7787	5621·2203	·4	287·1416	6561·1848	·1	308·1902	7558·3656
·7	266·0929	5634·5171	·5	287·4557	6575·5498	·2	308·5044	7573·7830
·8	266·4071	5647·8296	·6	287·7699	6589·9304	·3	308·8186	7589·2161
·9	266·7212	5661·1578	·7	288·0840	6604·3268	·4	309·1327	7604·6648
85·0	267·0354	5674·5017	·8	288·3982	6618·7388	·5	309·4469	7620·1293
·1	267·3495	5687·8614	·9	288·7124	6633·1666	·6	309·7610	7635·6095
·2	267·6637	5701·2367	92·0	289·0265	6647·6101	·7	310·0752	7651·1054
·3	267·9779	5714·6277	·1	289·3407	6662·0692	·8	310·3894	7666·6170
·4	268·2920	5728·0345	·2	289·6548	6676·5441	·9	310·7035	7682·1444
·5	268·6062	5741·4569	·3	289·9690	6691·0347	99·0	311·0177	7697·6874
·6	268·9203	5754·8951	·4	290·2832	6705·5410	·1	311·3318	7713·2461
·7	269·2345	5768·3490	·5	290·5973	6720·0630	·2	311·6460	7728·8206
·8	269·5486	5781·8185	·6	290·9115	6734·6008	·3	311·9602	7744·4107
·9	269·8628	5795·3038	·7	291·2256	6749·1542	·4	312·2743	7760·0166
86·0	270·1770	5808·8048	·8	291·5398	6763·7233	·5	312·5885	7775·6382
·1	270·4911	5822·3215	·9	291·8540	6778·3082	·6	312·9026	7791·2754
·2	270·8053	5835·8539	93·0	292·1681	6792·9087	·7	313·2168	7806·9284
·3	271·1194	5849·4020	·1	292·4823	6807·5250	·8	313·5309	7822·5971
·4	271·4336	5862·9659	·2	292·7964	6822·1569	·9	313·8451	7838·2815
·5	271·7478	5876·5454	·3	293·1106	6836·8046	100·0	314·1593	7853·9816
·6	272.0619	5890·1407						

Table XVII.

Properties of Sphere.

Surface of sphere $= 4\pi r^2 = \pi d^2 =$ area of 4 great circles $=$ circ. $\times$ diam. $=$ (circ.)$^2 \div \pi$.

Volume of sphere $= \frac{4}{3}\pi r^3 = \frac{1}{6}\pi d^3 = 0\cdot 01688 \times$ (circ.)^3 $= \frac{2}{3}$ circumscribing cylinder.

Segment of Sphere.—Surface $= 2\pi rh$, $h =$ height of segment.
$\qquad\qquad$ Volume $= \frac{2}{3}\pi (3r - h) h^2$, $r =$ rad. of sphere.
$\qquad\qquad\qquad\quad = \frac{\pi}{6}(3r_1^2 + h^2) h$, $r_1 =$ rad. of base of segment.

Sector of Sphere.—Surface as for segment. Volume $=$ segment $\pm$ cone.

Zone of Sphere.—Surface $= 2\pi rh$.
$\qquad\qquad$ Volume $= \dfrac{\pi}{2}\left(r_1^2 + \dfrac{h^2}{3} + r_2^2\right) h$.

(r_1, r_2 the radii of the parallel ends, h their distance apart).

Surface of middle zone $= 2\pi rh$.

Volume ditto $= \pi (r^2 - \frac{1}{12}h^2) h$.

APPENDIX A.

Table XVIII.

Table of Spheres.

Some errors of 1 in the last figure only.

Diam.	Surface.	Solidity.	Diam.	Surface.	Solidity.	Diam.	Surface.	Solidity.
	0·00077		2	12·566	4·1888		80·516	67·935
	0·00307	0·00002		12·962	4·3882		82·516	70·482
	0·00690	0·00005		13·364	4·5939		84·541	73·092
	0·01227	0·00013		13·772	4·8060		86·591	75·767
	0·02761	0·00043		14·186	5·0243		88·664	78·505
	0·04909	0·00102		14·607	5·2493		90·763	81·308
	0·07670	0·00200		15·033	5·4809		92·887	84·178
	0·11045	0·00345		15·466	5·7190		95·033	87·113
	0·15033	0·00548		15·904	5·9641		97·205	90·118
	0·19635	0·00818		16·349	6·2161		99·401	93·189
	0·24851	0·01165		16·800	6·4751		101·62	96·331
	0·30680	0·01598		17·258	6·7412		103·87	99·541
	0·37123	0·02127		17·721	7·0144		106·14	102·82
	0·44179	0·02761		18·190	7·2949		108·44	106·18
	0·51848	0·03511		18·666	7·5829		110·75	109·60
	0·60132	0·04385		19·147	7·8783	6	113·10	113·10
	0·69028	0·05393		19·635	8·1813		117·87	120·31
	0·78540	0·06545		20·129	8·4919		122·72	127·83
	0·88664	0·07850		20·629	8·8103		127·68	135·66
	0·99403	0·09319		21·135	9·1366		132·73	143·79
	1·1075	0·10960		21·648	9·4708		137·89	152·25
	1·2272	0·12783		22·166	9·8131		143·14	161·03
	1·3530	0·14798		22·691	10·164		148·49	170·14
	1·4849	0·17014		23·222	10·522	7	153·94	179·59
	1·6230	0·19442		23·758	10·889		159·49	189·39
	1·7671	0·22089		24·302	11·265		165·13	199·53
	1·9175	0·24967		24·850	11·649		170·87	210·03
	2·0739	0·28084		25·405	12·041		176·71	220·89
	2·2365	0·31451		25·967	12·443		182·66	232·13
	2·4053	0·35077		26·535	12·853		188·69	243·73
	2·5802	0·38971		27·109	13·272		194·83	255·72
	2·7611	0·43143		27·688	13·700	8	201·06	268·08
	2·9483	0·47603	3	28·274	14·137		207·39	280·85
1	3·1416	0·52360		29·465	15·039		213·82	294·01
	3·3410	0·57424		30·680	15·979		220·36	307·58
	3·5466	0·62804		31·919	16·957		226·98	321·56
	3·7583	0·68511		33·183	17·974		233·71	335·95
	3·9761	0·74551		34·472	19·031		240·53	350·77
	4·2000	0·80939		35·784	20·129		247·45	366·02
	4·4301	0·87681		37·122	21·268	9	254·47	381·70
	4·6664	0·94786		38·484	22·449		261·59	397·83
	4·9088	1·0227		39·872	23·674		268·81	414·41
	5·1573	1·1013		41·283	24·942		276·12	431·44
	5·4119	1·1839		42·719	26·254		283·53	448·92
	5·6728	1·2704		44·179	27·611		291·04	466·87
	5·9396	1·3611		45·664	29·016		298·65	485·31
	6·2126	1·4561		47·173	30·466		306·36	504·21
	6·4919	1·5553		48·708	31·965	10	314·16	523·60
	6·7771	1·6590	4	50·265	33·510		322·06	543·48
	7·0686	1·7671		51·848	35·106		330·06	563·86
	7·3663	1·8799		53·456	36·751		338·16	584·74
	7·6699	1·9974		55·089	38·448		346·36	606·13
	7·9798	2·1196		56·745	40·195		354·66	628·04
	8·2957	2·2468		58·427	41·994		363·05	650·46
	8·6180	2·3789		60·133	43·847		371·54	673·42
	8·9461	2·5161		61·863	45·752	11	380·13	696·91
	9·2805	2·6586		63·617	47·713		388·83	720·95
	9·6211	2·8062		65·397	49·729		397·61	745·31
	9·9678	2·9592		67·201	51·801		406·49	770·64
	10·321	3·1177		69·030	53·929		415·48	795·33
	10·680	3·2818		70·883	56·116		424·56	822·18
	11·044	3·4514		72·759	58·359		433·73	849·40
	11·416	3·6270		74·663	60·663		443·01	876·79
	11·793	3·8083		76·589	63·026	12	452·39	904·78
	12·177	3·9956	5	78·340	65·450		461·87	933·34

(M.B.)

Table of Spheres—(continued).

Diam.	Surface.	Solidity.	Diam.	Surface.	Solidity.	Diam.	Surface.	Solidity.
1/4	471·44	962·52	21	1385·5	4849·1	3/4	2780·5	13787
3/8	481·11	992·28	1/8	1402·0	4936·2	7/8	2804·0	13961
1/2	490·87	1022·7	1/4	1418·6	5024·3	30	2827·4	14137
5/8	500·73	1053·6	3/8	1435·4	5113·5	1/8	2851·1	14315
3/4	510·71	1085·3	1/2	1452·2	5203·7	1/4	2874·8	14494
7/8	520·77	1117·5	5/8	1469·2	5295·1	3/8	2898·7	14674
13	530·93	1150·3	3/4	1486·2	5387·4	1/2	2922·5	14856
1/8	541·19	1183·8	7/8	1503·3	5480·8	5/8	2946·6	15039
1/4	551·55	1218·0	22	1520·5	5575·3	3/4	2970·6	15224
3/8	562·00	1252·7	1/8	1537·9	5670·8	7/8	2994·9	15411
1/2	572·55	1288·3	1/4	1555·3	5767·6	31	3019·1	15599
5/8	583·20	1324·4	3/8	1572·8	5865·2	1/8	3043·6	15788
3/4	593·95	1361·2	1/2	1590·4	5964·1	1/4	3068·0	15979
7/8	604·80	1398·6	5/8	1608·2	6064·1	3/8	3092·7	16172
14	615·75	1436·8	3/4	1626·0	6165·2	1/2	3117·3	16366
1/8	626·80	1475·6	7/8	1643·9	6267·3	5/8	3142·1	16561
1/4	637·95	1515·1	23	1661·9	6370·6	3/4	3166·9	16758
3/8	649·17	1555·3	1/8	1680·0	6475·0	7/8	3192·0	16957
1/2	660·52	1596·3	1/4	1698·2	6580·6	32	3217·0	17157
5/8	671·95	1637·9	3/8	1716·5	6687·3	1/8	3242·2	17359
3/4	683·49	1680·3	1/2	1735·0	6795·2	1/4	3267·4	17563
7/8	695·13	1723·3	5/8	1753·5	6904·2	3/8	3292·9	17768
15	706·85	1767·2	3/4	1772·1	7014·3	1/2	3318·3	17974
1/8	718·69	1811·7	7/8	1790·8	7125·6	5/8	3343·9	18182
1/4	730·63	1857·0	24	1809·6	7238·2	3/4	3369·6	18392
3/8	742·65	1903·0	1/8	1828·5	7351·9	7/8	3395·4	18604
1/2	754·77	1949·8	1/4	1847·5	7466·7	33	3421·2	18817
5/8	767·00	1997·4	3/8	1866·6	7583·0	1/8	3447·3	19032
3/4	779·32	2045·7	1/2	1885·8	7700·1	1/4	3473·3	19248
7/8	791·73	2094·8	5/8	1905·1	7818·6	3/8	3499·5	19466
16	804·25	2144·7	3/4	1924·4	7938·3	1/2	3525·7	19685
1/8	816·85	2195·3	7/8	1943·9	8059·2	5/8	3552·1	19907
1/4	829·57	2246·8	25	1963·5	8181·3	3/4	3578·5	20129
3/8	842·40	2299·1	1/8	1983·2	8304·7	7/8	3605·1	20354
1/2	855·29	2352·1	1/4	2002·9	8429·2	34	3631·7	20580
5/8	868·31	2406·0	3/8	2022·9	8554·9	1/8	3658·5	20808
3/4	881·42	2460·6	1/2	2042·8	8682·0	1/4	3685·3	21037
7/8	894·63	2516·1	5/8	2062·9	8810·3	3/8	3712·3	21268
17	907·93	2572·4	3/4	2083·0	8939·9	1/2	3739·3	21501
1/8	921·33	2629·6	7/8	2103·4	9070·6	5/8	3766·5	21736
1/4	934·83	2687·6	26	2123·7	9202·8	3/4	3793·7	21972
3/8	948·43	2746·5	1/8	2144·2	9336·2	7/8	3821·1	22210
1/2	962·12	2806·2	1/4	2164·7	9470·8	35	3848·5	22449
5/8	975·91	2866·8	3/8	2185·5	9606·7	1/8	3876·1	22691
3/4	989·80	2928·2	1/2	2206·2	9744·0	1/4	3903·7	22934
7/8	1003·8	2990·5	5/8	2227·1	9882·5	3/8	3931·5	23179
18	1017·9	3053·6	3/4	2248·0	10022	1/2	3959·2	23425
1/8	1032·1	3117·7	7/8	2269·1	10164	5/8	3987·2	23674
1/4	1046·4	3182·6	27	2290·2	10306	3/4	4015·2	23924
3/8	1060·8	3248·5	1/8	2311·5	10450	7/8	4043·3	24176
1/2	1075·2	3315·3	1/4	2332·8	10595	36	4071·5	24429
5/8	1089·8	3382·9	3/8	2354·3	10741	1/8	4099·9	24685
3/4	1104·5	3451·5	1/2	2375·8	10889	1/4	4128·3	24942
7/8	1119·3	3521·0	5/8	2397·5	11038	3/8	4156·9	25201
19	1134·1	3591·4	3/4	2419·2	11189	1/2	4185·5	25461
1/8	1149·1	3662·8	7/8	2441·1	11341	5/8	4214·1	25724
1/4	1164·2	3735·0	28	2463·0	11494	3/4	4243·0	25988
3/8	1179·3	3808·2	1/8	2485·1	11649	7/8	4271·8	26254
1/2	1194·6	3882·5	1/4	2507·2	11805	37	4300·9	26522
5/8	1210·0	3957·6	3/8	2529·5	11962	1/8	4330·0	26792
3/4	1225·4	4033·7	1/2	2551·8	12121	1/4	4359·2	27063
7/8	1241·0	4110·8	5/8	2574·3	12281	3/8	4388·5	27337
20	1256·7	4188·8	3/4	2596·7	12443	1/2	4417·9	27612
1/8	1272·4	4267·8	7/8	2619·4	12606	5/8	4447·5	27889
1/4	1288·3	4347·8	29	2642·1	12770	3/4	4477·1	28168
3/8	1304·2	4428·8	1/8	2665·0	12936	7/8	4506·8	28449
1/2	1320·3	4510·9	1/4	2687·8	13103	38	4536·5	28731
5/8	1336·4	4593·9	3/8	2710·9	13272	1/8	4566·5	29016
3/4	1352·7	4677·9	1/2	2734·0	13442	1/4	4596·4	29302
7/8	1369·0	4763·0	5/8	2757·3	13614	3/8	4626·5	29590

APPENDIX A.

Table of Spheres—(continued).

Diam.	Surface.	Solidity.	Diam.	Surface.	Solidity.	Diam.	Surface.	Solidity.
	4656·7	29880		7050·9	55674		9940·2	93190
	4686·9	30173		7088·3	56115		9984·4	93812
	4717·3	30466		7125·6	56559		10029	94438
	4747·9	30762		7163·1	57006		10073	95066
39	4778·4	31059		7200·7	57455		10118	95697
	4809·0	31359	48	7238·3	57906		10163	96330
	4839·9	31661		7276·0	58360	57	10207	96967
	4870·8	31964		7313·9	58815		10252	97606
	4901·7	32270		7351·9	59274		10297	98248
	4932·7	32577		7389·9	59734		10342	98893
	4964·0	32886		7428·0	60197		10387	99541
	4995·3	33197		7466·3	60663		10432	100191
40	5026·5	33510		7504·5	61131		10478	100845
	5058·1	33826	49	7543·1	61601		10523	101501
	5089·6	34143		7581·6	62074	58	10568	102161
	5121·3	34462		7620·1	62549		10614	102823
	5153·1	34783		7658·9	63026		10660	103488
	5184·9	35106		7697·7	63506		10706	104155
	5216·8	35431		7736·7	63989		10751	104826
	5248·9	35758		7775·7	64474		10798	105499
41	5281·1	36087		7814·8	64961		10844	106175
	5313·3	36418	50	7854·0	65450		10890	106854
	5345·6	36751		7893·3	65941	59	10936	107536
	5378·1	37086		7932·8	66436		10983	108221
	5410·7	37423		7972·2	66934		11029	108909
	5443·3	37763		8011·8	67433		11076	109600
	5476·0	38104		8051·6	67935		11122	110294
	5508·9	38448		8091·4	68439		11169	110990
42	5541·9	38792		8131·3	68946		11216	111690
	5574·9	39140	51	8171·2	69456		11263	112392
	5608·0	39490		8211·4	69967	60	11310	113098
	5641·3	39841		8251·6	70482		11357	113806
	5674·5	40194		8292·0	70999		11404	114518
	5708·0	40551		8332·3	71519		11452	115232
	5741·5	40908		8372·8	72040		11499	115949
	5775·2	41268		8413·4	72567		11547	116669
43	5808·8	41630		8454·1	73092		11595	117392
	5842·7	41994	52	8494·8	73622		11642	118118
	5876·5	42360		8535·8	74154	61	11690	118847
	5910·7	42729		8576·8	74689		11738	119579
	5944·7	43099		8617·8	75226		11786	120315
	5978·9	43472		8658·9	75767		11834	121053
	6013·2	43846		8700·4	76309		11882	121794
	6047·7	44224		8741·7	76854		11931	122538
	6082·1	44602		8783·2	77401		11980	123286
44	6116·8	44984	53	8824·8	77952		12028	124036
	6151·5	45367		8866·4	78505	62	12076	124789
	6186·3	45753		8908·2	79060		12126	125545
	6221·2	46141		8950·1	79617		12174	126305
	6256·1	46530		8992·0	80178		12223	127067
	6291·2	46922		9034·1	80741		12272	127832
	6326·5	47317		9076·4	81308		12322	128601
45	6361·7	47713		9118·5	81876		12371	129373
	6397·2	48112		9160·8	82448		12420	130147
	6432·7	48513	54	9203·3	83021	63	12469	130925
	6468·3	48916		9246·0	83598		12519	131706
	6503·9	49321		9288·5	84177		12568	132490
	6539·7	49729		9331·2	84760		12618	133277
	6575·5	50139		9374·1	85344		12668	134067
	6611·6	50551		9417·2	85931		12718	134860
	6647·6	50965		9460·2	86521		12768	135657
46	6683·7	51382	55	9503·2	87114		12818	136456
	6720·0	51801		9546·5	87709	64	12868	137259
	6756·5	52222		9590·0	88307		12918	138065
	6792·9	52645		9633·3	88908		12969	138874
	6829·5	53071		9676·8	89511		13019	139686
	6866·1	53499		9720·6	90117		13070	140501
	6902·9	53929		9764·4	90726		13121	141320
47	6939·9	54362		9808·1	91338		13172	142142
	6976·8	54797	56	9852·0	91953		13222	142966
	7013·9	55234		9896·0	92570	65	13273	143794

(M.B.)

Table XVIII.—Table of Spheres (continued).

Diam.	Surface.	Solidity.	Diam.	Surface.	Solidity.	Diam.	Surface.	Solidity.
66	13324	144625	74 7/8	17146	211102	83	21448	295347
	13376	145460		17204	212175		21512	296691
	13427	146297	75 1/8	17262	213252		21578	298036
	13478	147138		17320	214333		21642	299388
	13530	147982		17379	215417		21708	300743
	13582	148828		17437	216505		21773	302100
	13633	149680		17496	217597		21839	303463
	13685	150533		17554	218693		21904	304831
	13737	151390		17613	219792		21970	306201
	13789	152251	75	17672	220894		22036	307576
	13841	153114		17731	222001		22102	308957
	13893	153980		17790	223111	84	22167	310340
	13946	154850		17849	224224		22234	311728
	13998	155724		17908	225341		22300	313118
	14050	156600		17968	226463		22366	314514
67	14103	157480		18027	227588		22432	315915
	14156	158363		18087	228716		22499	317318
	14208	159250	76	18146	229848		22565	318726
	14261	160139		18206	230984		22632	320140
	14314	161032		18266	232124	85	22698	321556
	14367	161927		18326	233267		22765	322977
	14420	162827		18386	234414		22832	324402
	14474	163731		18446	235566		22899	325831
68	14527	164637		18506	236719		22966	327264
	14580	165547		18566	237879		23034	328702
	14634	166460	77	18626	239041		23101	330142
	14688	167376		18687	240200		23168	331588
	14741	168295		18748	241376	86	23235	333039
	14795	169218		18809	242551		23303	334492
	14849	170145		18869	243728		23371	335951
	14903	171074		18930	244908		23439	337414
69	14957	172007		18992	246093		23506	333882
	15012	172944		19053	247283		23575	340352
	15066	173883	78	19114	248475		23643	341829
	15120	174828		19175	249672		23711	343307
	15175	175774		19237	250873	87	23779	344792
	15230	176723		19298	252077		23847	346281
	15284	177677		19360	253284		23916	347772
	15339	178635		19422	254496		23984	349269
70	15394	179595		19483	255713		24053	350771
	15449	180559		19545	256932		24122	352277
	15504	181525	79	19607	258155		24191	353785
	15560	182497		19669	259383		24260	355301
	15615	183471		19732	260613	88	24328	356819
	15670	184449		19794	261848		24398	358342
	15726	185430		19856	263088		24467	359860
	15782	186414		19919	264330		24536	361400
71	15837	187402		19981	265577		24606	362935
	15893	188394		20044	266829		24676	364476
	15949	189389	80	20106	268083		24745	366019
	16005	190387		20170	269342		24815	367568
	16061	191389		20232	270604	89	24885	369122
	16117	192395		20296	271871		24955	370678
	16174	193404		20358	273141		25025	372240
	16230	194417		20422	274416		25095	373806
72	16286	195433		20485	275694		25165	375378
	16343	196453		20549	276977		25236	376954
	16400	197476	81	20612	278263		25306	378531
	16456	198502		20676	279553		25376	380115
	16513	199532		20740	280847	90	25447	381704
	16570	200566		20804	282145		25518	383297
	16628	201604		20887	283447		25589	384894
	16685	202645		20932	284754		25660	386496
73	16742	203689		20996	286064		25730	388102
	16799	204737		21060	287378		25802	389711
	16857	205789	82	21124	288696		25873	391327
	16914	206844		21189	290019		25944	392945
	16972	207903		21253	291345	91	26016	394570
	17030	208966		21318	292674		26087	396197
	17088	210032		21382	294010		26159	397831

APPENDIX A.

TABLE XVIII.—Table of Spheres (continued).

Diam.	Surface.	Solidity.	Diam.	Surface.	Solidity.	Diam.	Surface.	Solidity.
	26230	399468		27981	440118		29712	481579
	26302	401109		28055	441871		29788	483438
	26374	402756		28130	443625		29865	485302
	26446	404406		28204	445387		29942	487171
	26518	406060		28278	447151		30018	489045
92	26590	407721	95	28353	448920		30095	490924
	26663	409384		28428	450695	98	30172	492808
	26735	411054		28503	452475		30249	494695
	26808	412726		28577	454259		30326	496588
	26880	414405		28652	456047		30404	498486
	26953	416086		28727	457839		30481	500388
	27026	417774		28802	459638		30558	502296
	27099	419464		28878	461439		30636	504208
93	27172	421161	96	28953	463248		30713	506125
	27245	422862		29028	465059	99	30791	508047
	27318	424567		29104	466875		30869	509975
	27391	426277		29180	468697		30947	511906
	27464	427991		29255	470524		31025	513843
	27538	429710		29331	472354		31103	515785
	27612	431433		29407	474189		31181	517730
	27686	433160		29483	476029		31259	519682
94	27759	434894	97	29559	477874		31338	521638
	27833	436630		29636	479725	100	31416	523598
	27907	438373						

MANUAL OF MILITARY BALLOONING.

TABLE XIX.
Table of Logarithms of Numbers 0 to 1,000.*

No.	0.	1.	2.	3.	4.	5.	6.	7.	8.	9.	Prop.
0	0	00000	30103	47712	60206	69897	77815	84510	90309	95424	
10	00000	00432	00860	01283	01703	02118	02530	02938	03342	03742	415
11	04139	04532	04921	05307	05690	06069	06445	06818	07188	07554	379
12	07918	08278	08636	08990	09342	09691	10037	10380	10721	11059	349
13	11394	11727	12057	12385	12710	13033	13353	13672	13987	14301	323
14	14613	14921	15228	15533	15836	16136	16435	16731	17026	17318	300
15	17609	17897	18184	18469	18752	19033	19312	19590	19865	20139	281
16	20412	20682	20951	21218	21484	21748	22010	22271	22530	22788	264
17	23045	23299	23552	23804	24054	24303	24551	24797	25042	25285	249
18	25527	25767	26007	26245	26481	26717	26951	27184	27415	27646	236
19	27875	28103	28330	28555	28780	29003	29225	29446	29666	29885	223
20	30103	30319	30535	30749	30963	31175	31386	31597	31806	32014	212
21	32222	32428	32633	32838	33041	33243	33445	33646	33845	34044	202
22	34242	34439	34635	34830	35024	35218	35410	35602	35793	35983	194
23	36173	36361	36548	36735	36921	37106	37291	37474	37657	37839	185
24	38021	38201	38381	38560	38739	38916	39093	39269	39445	39619	177
25	39794	39967	40140	40312	40483	40654	40824	40993	41162	41330	171
26	41497	41664	41830	41995	42160	42324	42488	42651	42813	42975	164
27	43136	43296	43456	43616	43775	43933	44090	44248	44404	44560	158
28	44716	44870	45024	45178	45331	45484	45636	45788	45939	46089	153
29	46240	46389	46538	46686	46834	46982	47129	47275	47421	47567	148
30	47712	47856	48000	48144	48287	48430	48572	48713	48855	48995	143
31	49136	49276	49415	49554	49693	49831	49968	50105	50242	50379	138
32	50515	50650	50785	50920	51054	51188	51321	51454	51587	51719	134
33	51851	51982	52113	52244	52374	52504	52633	52763	52891	53020	130
34	53148	53275	53402	53529	53655	53781	53907	54033	54157	54282	126
35	54407	54530	54654	54777	54900	55022	55145	55266	55388	55509	122
36	55630	55750	55870	55990	56110	56229	56348	56466	56584	56702	119
37	56820	56937	57054	57170	57287	57403	57518	57634	57749	57863	116
38	57978	58092	58206	58319	58433	58546	58658	58771	58883	58995	113
39	59106	59217	59328	59439	59549	59659	59769	59879	59988	60097	110
40	60206	60314	60422	60530	60638	60745	60852	60959	61066	61172	107
41	61278	61384	61489	61595	61700	61804	61909	62013	62118	62221	104
42	62325	62428	62531	62634	62736	62838	62941	63042	63144	63245	102
43	63347	63447	63548	63648	63749	63848	63948	64048	64147	64246	99
44	64345	64443	64542	64640	64738	64836	64933	65030	65127	65224	98
45	65321	65417	65513	65609	65705	65801	65896	65991	66086	66181	96
46	66276	66370	66464	66558	66651	66745	66838	66931	67024	67117	94
47	67210	67302	67394	67486	67577	67669	67760	67851	67942	68033	92
48	68124	68214	68304	68394	68484	68574	68663	68752	68842	68930	90
49	69020	69108	69196	69284	69372	69460	69548	69635	69722	69810	88
50	69897	69983	70070	70156	70243	70329	70415	70500	70586	70671	86
51	70757	70842	70927	71011	71096	71180	71265	71349	71433	71516	84
52	71600	71683	71767	71850	71933	72015	72098	72181	72263	72345	82
53	72428	72509	72591	72672	72754	72835	72916	72997	73078	73158	81
54	73239	73319	73399	73480	73559	73639	73719	73798	73878	73957	80
55	74036	74115	74193	74272	74351	74429	74507	74585	74663	74741	78
56	74818	74896	74973	75050	75127	75204	75281	75358	75434	75511	77
57	75587	75663	75739	75815	75891	75966	76042	76117	76192	76267	75
58	76342	76417	76492	76566	76641	76715	76789	76863	76937	77011	74
59	77085	77158	77232	77305	77378	77451	77524	77597	77670	77742	73
60	77815	77887	77959	78031	78103	78175	78247	78318	78390	78461	72
61	78533	78604	78675	78746	78816	78887	78958	79028	79098	79169	71
62	79239	79309	79379	79448	79518	79588	79657	79726	79796	79865	70
63	79934	80002	80071	80140	80208	80277	80345	80413	80482	80550	69
64	80618	80685	80753	80821	80888	80956	81023	81090	81157	81224	68
65	81291	81358	81424	81491	81557	81624	81690	81756	81822	81888	67
66	81954	82020	82085	82151	82216	82282	82347	82412	82477	82542	66
67	82607	82672	82736	82801	82866	82930	82994	83058	83123	83187	65
68	83250	83314	83378	83442	83505	83569	83632	83695	83758	83821	64
69	83884	83947	84010	84073	84136	84198	84260	84323	84385	84447	63
70	84509	84571	84633	84695	84757	84818	84880	84941	85003	85064	62

* Each log is supposed to have the decimal sign before it.

APPENDIX B.

TABLE XIX.—Table of Logarithms of Numbers 0 to 1,000*
(continued).

No.	0.	1.	2.	3.	4.	5.	6.	7.	8.	9.	Prop.
71	85125	85187	85248	85309	85369	85430	85491	85551	85612	85672	61
72	85733	85793	85853	85913	85973	86033	86093	86153	86213	86272	60
73	86332	86391	86451	86510	86569	86628	86687	86746	86805	86864	59
74	86923	86981	87040	87098	87157	87215	87273	87332	87390	87448	58
75	87506	87564	87621	87679	87737	87794	87852	87909	87966	88024	57
76	88081	88138	88195	88252	88309	88366	88422	88479	88536	88592	56
77	88649	88705	88761	88818	88874	88930	88986	89042	89098	89153	56
78	89209	89265	89320	89376	89431	89487	89542	89597	89652	89707	55
79	89762	89817	89872	89927	89982	90036	90091	90145	90200	90254	54
80	90309	90363	90417	90471	90525	90579	90633	90687	90741	90794	54
81	90848	90902	90955	91009	91062	91115	91169	91222	91275	91328	53
82	91381	91434	91487	91540	91592	91645	91698	91750	91803	91855	53
83	91907	91960	92012	92064	92116	92168	92220	92272	92324	92376	52
84	92427	92479	92531	92582	92634	92685	92737	92788	92839	92890	51
85	92941	92993	93044	93095	93146	93196	93247	93298	93348	93399	51
86	93449	93500	93550	93601	93651	93701	93751	93802	93852	93902	50
87	93951	94001	94051	94101	94151	94200	94250	94300	94349	94398	49
88	94448	94497	94546	94596	94645	94694	94743	94792	94841	94890	49
89	94939	94987	95036	95085	95133	95182	95230	95279	95327	95376	48
90	95424	95472	95520	95568	95616	95664	95712	95760	95808	95856	48
91	95904	95951	95999	96047	96094	96142	96189	96236	96284	96331	48
92	96378	96426	96473	96520	96567	96614	96661	96708	96754	96801	47
93	96848	96895	96941	96988	97034	97081	97127	97174	97220	97266	47
94	97312	97359	97405	97451	97497	97543	97589	97635	97680	97726	46
95	97772	97818	97863	97909	97954	98000	98045	98091	98136	98181	46
96	98227	98272	98317	98362	98407	98452	98497	98542	98587	98632	45
97	98677	98721	98766	98811	98855	98900	98945	98989	99033	99078	45
98	99122	99166	99211	99255	99299	99343	99387	99431	99475	99519	44
99	99563	99607	99651	99694	99738	99782	99825	99869	99913	99956	44

* Each log is supposed to have the decimal sign · before it.

The log of 2870 is 3·45788 | The log of 0·287 is − 1·45788
  ,, ,,   287 is 2·45788 | ,, ,,   0·028 is − 2·44716
  ,, ,,   28·7 is 1·45788 | ,, ,,   0·002 is − 3·30103
  ,, ,,   2·87 is 0·45788 | ,, ,,   0·0002 is − 4·30103

What is the log of 2873?

Here, log of 2870 = 3·45788
And prop 153 × 3 =   459
_____
3·458339

To find roots, divide the log (with its index) of the given number by that number which expresses the kind of root. The quotient will be the log of the required root.

Example. What is the cube root of 2870?

Here, the log of 2870, with its index, is 3·45788. And $\frac{3·45788}{3}$ = 1·15263.

Hence the cube root is 14·2.

The Hyperbolic or Napierian logarithm is the common log of the table multiplied by 2·3025851.

APPENDIX B.

TABLES FOR FACILITATING CALCULATIONS RESPECTING THE USE OF BALLOONS.

Compiled by Lieut.-Colonel C. M. WATSON, C.M.G., R.E.

NOTE.

THESE Tables have been compiled to facilitate the calculations which have constantly to be made in the conducting of Balloon operations, and to explain the manner in which the volumes of gases vary under different meteorological conditions.

Neglect of these conditions, and an unscientific use of gas will lead to a useless waste of it, which is especially to be avoided in Military Ballooning. I have observed that errors frequently arise from the common practice of speaking of gas in cubic feet, which is a measure quite devoid of meaning unless the barometric pressure and temperature are also given. It would be better if gas could always be described by weight and specific gravity, as the quantity so described is unvariable and independent of atmospheric influence. For example, 1 lb. of hydrogen gas of specific gravity 0·07 will always be able to lift a weight of 13·28 lbs., no matter what the pressure and temperature may be. But 100 cubic feet of hydrogen at 30 inches and 60° become 120 feet at 25 inches and 60°, and whereas 100 feet would lift 7·13 lbs. at the former temperature and pressure, it will only lift 5·94 lbs. at the latter.

It should be noted that the term "specific gravity" is used throughout as meaning "the weight of any volume of gas at "any temperature and pressure, divided by the weight of the "*same* volume of air at the *same* temperature and pressure."

It is obvious that the specific gravity of any gas as thus defined remains constant under all conditions of temperature and pressure. Also, that the specific gravity of air is always 1.

The air and gases are assumed to be dry. No allowance has been made in the tables or in the calculations for the aqueous vapour which they may contain.

EXPLANATION OF THE TABLES.

TABLE I.

This table gives some useful data with regard to gases of different specific gravities. If it is desired to convert the figures

APPENDIX B.

in column 2 to the figures corresponding to pressures and temperatures other than 30·0 inches and 60°, this can be done by multiplying them by the numbers given in Table II.

Likewise, if it is desired to convert the figures in columns 4 and 5 to the figures corresponding to other pressures and temperatures, they should be multiplied by the numbers given in Table III.

Example. What volume will 1 lb. weight of gas of specific gravity 0·075 occupy at 24·0 inches and 40°?

From Table I, column 2, it appears that the volume of 1 lb. of gas of specific gravity 0·075 at 30·0 inches and 60° is 173·733 cubic feet. From Table II we find that the multiplier for 24·0 inches and 40° is 1·2017—

$$173{\cdot}733 \times 1{\cdot}2017 = 208{\cdot}775,$$

therefore 1 lb. of gas of specific gravity 0·075 will occupy 208·775 cubic feet at pressure 24·0 inches and temperature 40°.

If it is desired to deal with gases of which the specific gravities are not given in Table I, the figures for each column can be calculated in the following manner:—

Column 2.—Divide the figure at the head of the column by the specific gravity of the gas.

Column 3.—Calculate the reciprocal of the specific gravity of the gas, and subtract 1 from it.

Column 4.—Multiply the figure at the head of the column by the specific gravity of the gas.

Column 5.—Subtract the figure in column 4 from the figure at the head of column 4.

Example. Calculate the figures for columns 2, 3, 4, 5 for a gas of specific gravity 0·116.

Column 2. $\dfrac{13{\cdot}036}{0{\cdot}116} = 112{\cdot}38$ cubic feet.

Column 3. $\dfrac{1}{0{\cdot}116} - 1 = 7{\cdot}6207$ lbs.

Column 4. $767{\cdot}08 \times 0{\cdot}116 = 88{\cdot}98$ lbs.

Column 5. $767{\cdot}08 - 88{\cdot}98 = 678{\cdot}10$ lbs.

TABLE II.

By this table any volume of gas at pressure 30 inches and temperature 60° can be reduced to the volume which the same weight of gas will occupy at any other pressure and temperature.

Example. What will be the volume of 50 lbs. of hydrogen of specific gravity 0·070 at 23 inches and 50°?

We find by Table I, column 2, that 1 lb. of gas of specific

MANUAL OF MILITARY BALLOONING.

gravity 0·070 will occupy 186·23 cubic feet at 30 inches and 60°. And the multiplier in Table II for 23 inches and 50° is 1·2792.

Multiplying out—

$$50 \times 186·23 \times 1·2792 = 11911·27,$$

therefore, 50 lbs. of gas of specific gravity 0·070 will have a volume of 11,911¼ cubic feet at 23 inches and 60°.

If it is required to find the multiplier for a pressure and temperature between those entered in the table, this can be easily done by interpolation.

Example. What will be the volume at 24·7 inches and 55° of a quantity of gas which occupies 8,000 cubic feet at 30 inches and 60°.

By Table II—
    The multiplier for 24·5 inches and 50° is 1·2008.
            „   „ 25·0 inches and 50° is 1·1768.
    Therefore „   „ 24·7 inches and 50° is 1·1912.
    The multiplier for 24·5 inches and 60° is 1·2244.
            „   „ 25·0 inches and 60° is 1·1999.
    Therefore „   „ 24·7 inches and 60° is 1·2146.

Combining the results—
    The multiplier for 24·7 inches and 60° is 1·2146.
            „   „ 24·7 inches and 50° is 1·1912.
    Therefore „   „ 24·7 inches and 55° is 1·2029.

$$8,000 \times 1·2029 = 9623·20.$$

Therefore 8,000 cubic feet of gas at 30 inches and 60° will become 9,623¼ cubic feet at 24·7 inches and 55°.

The above method of obtaining the intermediate multipliers is not absolutely accurate, but is sufficiently near for practical purposes. For example, in the instance given above, the real multiplier would be 1·20283.

Table III.

This table is the converse of Table II, and serves to convert a volume of gas at any temperature and pressure to the volume which the same quantity of gas will occupy at 30 inches and 60°.

Example. A 20,000 feet balloon is just full at 22·5 inches and 30°. What volume of gas will be in it at 30 inches and 60°?

In Table III the multiplier corresponding to 22·5 inches and 30° is 0·7959—

$$20,000 \times 0·7959 = 15,918.$$

APPENDIX B.

Therefore the gas which occupied 20,000 cubic feet at 22·5 inches and 30° will have contracted to 15,918 cubic feet at 30 inches and 60°.

By combining Tables II and III, a volume of gas at any temperature and pressure can be reduced to the volume which the same weight of gas will occupy at any other temperature and pressure.

Example. A certain weight of gas has a volume of 15,000 cubic feet at 29·5 inches and 70°. What volume will it occupy at 23 inches and 40°?

From Table III, the multiplier for 29·5 inches and 70° is 0·9649.

From Table II, the multiplier for 23 inches and 40° is 1·2540.

$$15,000 \times 0·9649 \times 1·2540 = 18149·76.$$

Therefore 15,000 cubic feet of gas at 29·5 inches and 70° will become 18149·76 cubic feet at 23 inches and 40°.

It should be remembered that Table III reduces the gas *to* what it occupies at 30 inches and 60°, while Table II reduces it *from* 30 inches and 60°.

TABLE IV.

This table gives the weight of 10,000 cubic feet of air at different pressures and temperatures, and from these data the lifting power of 10,000 cubic feet of any gas at any pressure and temperature can be obtained.

N.B.—Lifting power may be defined as the difference in weight of the gas in the balloon and of the displaced air.

Example. What is the lifting power of 10,000 cubic feet of coal gas of sp. gr. 0·4 at 24 inches and 40° F.?

From the table the weight of 10,000 cubic feet of air at 24 inches and 40° = 638·3.

Therefore the weight of an equal volume of a gas whose specific gravity is $0·4 = 638·3 \times 0·4 = 255·32$.

Therefore the lifting power of the gas $= 638·3 - 255·32 = 382·98$ lbs.

TABLE V.

Table V gives the lifting power of hydrogen at different pressures and temperatures.

It is calculated for 10,000 cubic feet, but can be used for any other quantity.

Example. What is the lifting power of 7,500 cubic feet of hydrogen at 25 inches and 60°?

From the table, the lifting power of 10,000 feet of hydrogen at 25 inches and 60° is 595·1 lbs—

$$595·1 \times 0·75 = 446·32.$$

Therefore the lifting power of 7,500 cubic feet of hydrogen at 25 inches and 60° is 446·32 lbs.

Table VI.

In this table is given the height of barometer due to different altitudes, the temperature being the same at both stations, on the assumption that 30 inches corresponds to altitude 0.

In column 3 is shown the volume which a quantity of gas, measuring 10,000 cubic feet at level 0, will fill at the different altitudes given in column 1, assuming that the temperature remains constant.

It is self-evident that the figures in this table can only be taken as approximately correct, as 30·0 inches does not always correspond to 0 altitude.

Table VII.

The difference in altitude between two stations can be ascertained with very considerable accuracy by means of the figures in this table.

The observations at both stations should, if possible, be taken simultaneously.

The table is only calculated to tenths of an inch on the barometric scale, and the change due to hundredths must be taken from column 3 and subtracted.

The formula to be used is $(F - f)\left(1 + \dfrac{T + t - 100}{1000}\right)$ where F is the figure in column 2, corresponding to the barometric height at the upper station—

T the temperature at the upper station,

and $f, t,$ are the same for the lower station.

Example. At station A the barometer is 29·98 inches, and thermometer 65°. At station B the barometer is 22·68 inches and thermometer 45°. What is the difference in altitude?

By Table VII—

For barometer 22·60 inches.	Figure in column 2	..	8613·0
Subtract 8 × 12·1	.. . ..	..	96·8
			8516·2
For barometer 29·90 inches.	Figure in column 2	..	984·0
Subtract 8 × 9·1	.. .. ..	..	72·8
			911·2

Diagram,

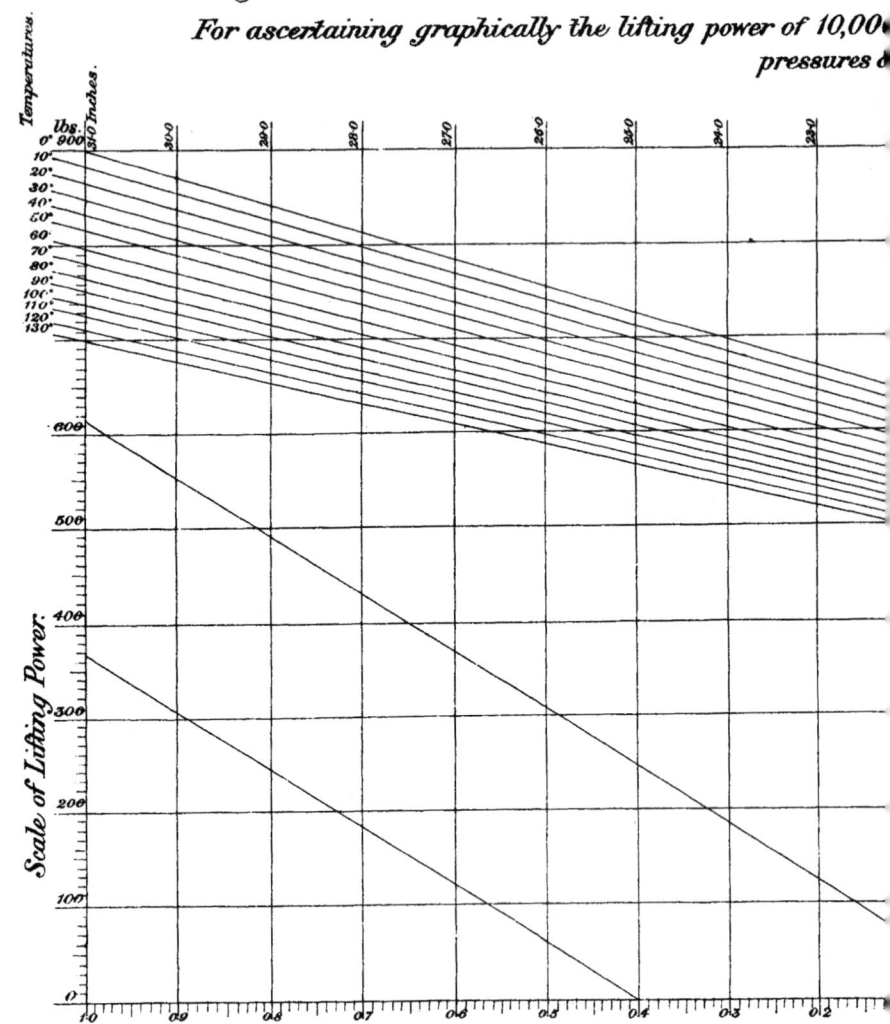

For ascertaining graphically the lifting power of 10,000 pressures

cubic feet of gas of given sp. gravity at different temperatures.

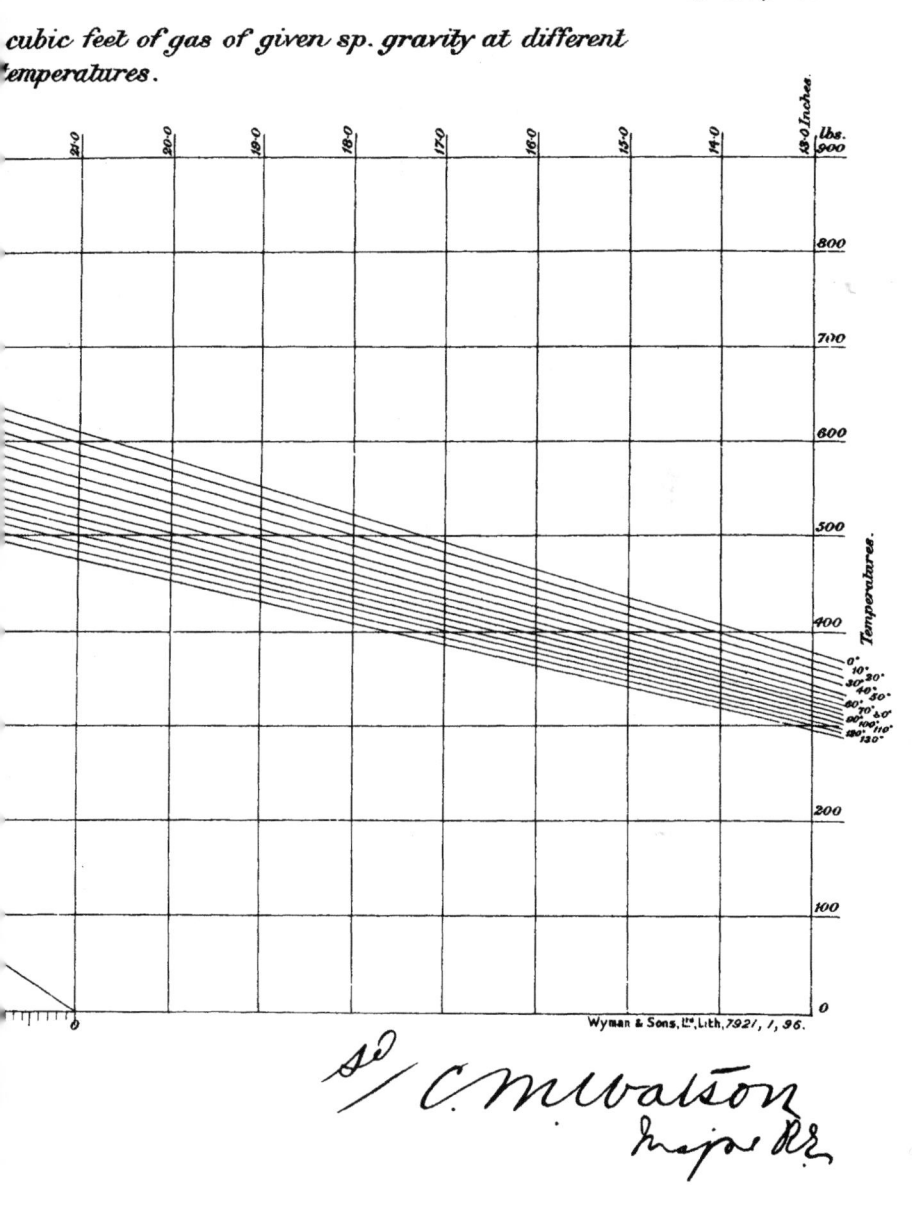

APPENDIX B.

By formula—

Difference of altitude $= (8516\cdot2 - 911\cdot2)\left(1 + \dfrac{65° + 45° - 100}{1000}\right)$
$= 7691\cdot05.$

Therefore B is 7,691 feet above A.

Altitudes can be ascertained much more accurately by the use of the above formula and Table VII, than by the scale of heights which are sometimes marked on aneroid barometers.

DIAGRAM.

By this diagram the lifting power of 10,000 cubic feet of gas of any specific gravity at different temperatures and pressures can be ascertained graphically with facility.

The vertical lines give the height of the barometer in inches.

The horizontal lines represent lbs.

The diagonal lines indicate different temperatures.

The point at which any of the three kinds of lines meet gives the weight of 10,000 cubic feet of air at the pressures and temperatures indicated.

To ascertain the lifting power of 10,000 cubic feet of gas of a given specific gravity at any pressure and temperature, the following rules should be followed:—

1. Ascertain the weight of 10,000 cubic feet of air at the given pressure and temperature.

2. Set off the weight so found on the "scale of lifting power."

3. Join the point so found with zero on the "scale of specific gravity."

4. Mark the specific gravity of the gas on the "scale of specific gravity."

5. Draw a line from the point so found, parallel to the line drawn in 3.

6. The point where the line so drawn cuts the "scale of lifting power," will give the lifting power of the gas for the pressure and temperature specified.

Table I.

Data respecting Gases of Different Specific Gravities.

	Column 1. Specific gravity of gas.	Column 2. Volume of 1 lb. weight of gas at 30 inches and 60° F.	Column 3. Lifting power of 1 lb. weight of gas at any pressure and temperature.	Column 4. Weight of 10,000 cubic feet of gas at 30 inches and 60° F.	Column 5. Lifting power of 10,000 cubic feet of gas at 30 inches and 60° F.
		Cubic feet.	lbs.	lbs.	lbs.
Air.	1·000	13·036	—	767·08	—
	0·900	14·485	0·111	690·37	76·71
	0·800	16·295	0·250	613·66	153·42
	0·700	18·623	0·428	536·96	230·12
	0·600	21·727	0·666	460·25	306·83
	0·500	26·073	1·000	383·54	383·54
	0·450	28·970	1·222	335·19	431·89
	0·400	32·591	1·500	306·83	460·25
	0·350	37·247	1·857	268·48	498·60
	0·300	43·455	2·333	230·12	536·96
	0·250	52·145	3·000	191·77	575·31
	0·200	65·182	4·000	153·42	613·66
	0·175	74·491	4·712	134·23	632·85
	0·150	86·909	5·666	115·06	652·02
	0·145	89·903	5·897	111·23	645·85
	0·140	93·114	6·143	107·39	659·69
	0·135	96·566	6·333	103·55	663·53
	0·130	100·280	6·692	99·72	667·36
	0·128	101·847	6·812	98·19	668·89
	0·126	103·463	6·937	96·65	670·43
	0·124	105·132	7·064	95·12	671·96
	0·122	106·855	7·197	93·58	673·50
	0·120	108·636	7·333	92·05	675·03
	0·118	110·478	7·474	90·51	676·57
	0·116	112·383	7·621	88·98	678·10
	0·114	114·354	7·772	87·45	679·63
	0·112	116·396	7·928	85·91	681·17
	0·110	118·512	8·091	84·38	682·70
	0·108	120·707	8·259	82·84	684·24
	0·106	122·985	8·432	81·31	685·77
	0·104	125·350	8·615	79·78	687·30
	0·102	127·808	8·804	78·24	688·84
	0·100	130·364	9·000	76·71	690·37
	0·098	133·021	9·204	75·17	691·91
	0·096	135·792	9·416	73·64	693·44
	0·094	138·681	9·638	72·11	694·97
	0·092	141·697	9·870	70·57	696·51
	0·090	144·849	10·111	69·04	698·04
	0·088	148·136	10·364	67·50	699·58
	0·086	151·581	10·628	65·97	701·11
	0·084	155·191	10·905	64·43	702·65
	0·082	158·976	11·195	62·90	704·18

APPENDIX B.

TABLE I—(*continued*).

Column 1. Specific gravity of gas.	Column 2. Volume of 1 lb. weight of gas at 30 inches and 60° F.	Column 3. Lifting power of 1 lb. weight of gas at any pressure and temperature.	Column 4. Weight of 10,000 cubic feet of gas at 30 inches and 60° F.	Column 5. Lifting power of 10,000 cubic feet of gas at 30 inches and 60° F.	
	Cubic feet.	lbs.	lbs.	lbs.	
0·080	162·955	11·500	61·37	705·71	
0·079	165·012	11·658	60·60	706·48	
0·078	167·136	11·820	59·83	707·25	
0·077	169·296	11·987	59·06	708·02	
0·076	171·526	12·158	58·30	708·78	
0·075	173·733	12·333	57·53	709·55	
0·074	176·163	12·513	56·76	710·32	
0·073	178·576	12·698	55·99	711·09	
0·072	181·055	12·888	55·23	711·85	
0·071	183·606	13·084	54·46	712·62	
0·070	186·234	13·280	53·70	713·38	
0·0692	188·383	13·451	53·08	714·00	Pure hydrogen.

TABLE II.

Multipliers (1) *to convert a volume of gas at Pressure 30 inches and Temperature 60° to the volume which the same weight of gas will occupy at other Pressures and Temperatures;* (2) *to convert the weight or lifting power of a given volume of gas at different Pressures and Temperatures to the weight or lifting power of the same volume at 30 inches and 60°.*

Temperature.	0°	10°	20°	30°	40°	50°	60°	70°	80°	90°	100°	110°	120°	130°
31·0	0·8558	0·8745	0·8931	0·9118	0·9304	0·9491	0·9677	0·9863	1·0050	1·0236	1·0423	1·0609	1·0796	1·0982
30·5	0·8699	0·8888	0·9078	0·9267	0·9457	0·9646	0·9836	1·0025	1·0215	1·0404	1·0594	1·0783	1·0973	1·1162
30·0	0·8844	0·9037	0·9229	0·9422	0·9615	0·9807	1·0000	1·0193	1·0385	1·0578	1·0771	1·0963	1·1156	1·1349
29·5	0·8993	0·9187	0·9385	0·9581	0·9777	0·9973	1·0169	1·0365	1·0561	1·0757	1·0953	1·1149	1·1345	1·1541
29·0	0·9148	0·9317	0·9547	0·9746	0·9945	1·0144	1·0344	1·0543	1·0743	1·0942	1·1141	1·1340	1·1540	1·1739
28·5	0·9308	0·9511	0·9714	0·9917	1·0119	1·0322	1·0525	1·0728	1·0931	1·1133	1·1336	1·1539	1·1742	1·1945
28·0	0·9474	0·9681	0·9887	1·0094	1·0300	1·0506	1·0713	1·0919	1·1126	1·1332	1·1539	1·1745	1·1951	1·2156
27·5	0·9647	0·9857	1·0067	1·0277	1·0487	1·0698	1·0908	1·1118	1·1328	1·1538	1·1749	1·1959	1·2169	1·2379
27·0	0·9826	1·0040	1·0254	1·0468	1·0682	1·0896	1·1110	1·1324	1·1538	1·1752	1·1966	1·2180	1·2394	1·2608
26·5	1·0011	1·0229	1·0447	1·0665	1·0884	1·1102	1·1320	1·1538	1·1756	1·1974	1·2192	1·2410	1·2629	1·2847
26·0	1·0204	1·0426	1·0649	1·0871	1·1093	1·1316	1·1538	1·1760	1·1985	1·2207	1·2429	1·2651	1·2874	1·3096
25·5	1·0404	1·0631	1·0857	1·1084	1·1311	1·1537	1·1764	1·1991	1·2217	1·2444	1·2671	1·2897	1·3124	1·3351
25·0	1·0612	1·0843	1·1074	1·1305	1·1537	1·1768	1·1999	1·2230	1·2461	1·2692	1·2924	1·3155	1·3386	1·3617
24·5	1·0828	1·1064	1·1300	1·1536	1·1772	1·2008	1·2244	1·2480	1·2716	1·2952	1·3188	1·3423	1·3659	1·3895
24·0	1·1054	1·1295	1·1535	1·1776	1·2017	1·2258	1·2499	1·2740	1·2981	1·3221	1·3462	1·3703	1·3944	1·4185
23·5	1·1289	1·1535	1·1781	1·2027	1·2273	1·2519	1·2765	1·3011	1·3257	1·3503	1·3749	1·3995	1·4241	1·4487
23·0	1·1535	1·1786	1·2038	1·2289	1·2540	1·2792	1·3043	1·3294	1·3546	1·3797	1·4048	1·4299	1·4551	1·4802
22·5	1·1792	1·2048	1·2305	1·2562	1·2819	1·3076	1·3333	1·3590	1·3847	1·4104	1·4361	1·4617	1·4874	1·5131
22·0	1·2059	1·2322	1·2585	1·2848	1·3110	1·3373	1·3636	1·3898	1·4161	1·4424	1·4687	1·4950	1·5212	1·5475
21·5	1·2340	1·2609	1·2878	1·3146	1·3415	1·3684	1·3953	1·4222	1·4491	1·4759	1·5028	1·5297	1·5566	1·5835
21·0	1·2633	1·2909	1·3184	1·3459	1·3734	1·4010	1·4285	1·4560	1·4835	1·5111	1·5386	1·5661	1·5936	1·6212
20·5	1·2912	1·3224	1·3506	1·3788	1·4070	1·4352	1·4634	1·4916	1·5198	1·5480	1·5762	1·6044	1·6326	1·6608
20·0	1·3266	1·3555	1·3844	1·4133	1·4422	1·4711	1·5000	1·5289	1·5578	1·5867	1·6156	1·6445	1·6734	1·7023

Height of Barometer in inches (corrected).

APPENDIX B.

TABLE III.

Multipliers (1) to convert a volume of gas at different Temperatures and Pressures to the volume which the same weight of gas would occupy at Pressure 30 inches and Temperature 60°; (2) to convert the weight or lifting power of a given volume of gas at 30 inches and 60° to the weight or lifting power of the same volume of gas at other Pressures and Temperatures.

Temperature.	0°	10°	20°	30°	40°	50°	60°	70°	80°	90°	100°	110°	120°	130°
31·0	1·1655	1·1435	1·1196	1·0967	1·0748	1·0536	1·0334	1·0139	0·9950	0·9759	0·9594	0·9426	0·9263	0·9106
30·5	1·1496	1·1251	1·1015	1·0790	1·0574	1·0366	1·0167	0·9976	0·9790	0·9611	0·9439	0·9274	0·9114	0·8959
30·0	1·1307	1·1067	1·0835	1·0613	1·0401	1·0196	1·0000	0·9812	0·9630	0·9453	0·9–85	0·9122	0·8964	0·8812
29·5	1·1120	1·0882	1·0654	1·0436	1·0228	1·0026	0·9834	0·9649	0·9469	0·9296	0·9130	0·8969	0·8815	0·8665
29·0	1·0931	1·0698	1·0474	1·0259	1·0054	0·9856	0·9667	0·9485	0·9309	0·9138	0·8975	0·8818	0·8665	0·8518
28·5	1·0743	1·0513	1·0293	1·0082	0·9881	0·9686	0·9500	0·9321	0·9143	0·8931	0·8821	0·8666	0·8516	0·8371
28·0	1·0555	1·0329	1·0112	0·9905	0·9703	0·9516	0·9334	0·9158	0·8988	0·8823	0·8666	0·8514	0·8367	0·8224
27·5	1·0366	1·0144	0·9932	0·9729	0·9534	0·9346	0·9167	0·8994	0·8827	0·8666	0·8511	0·8362	0·8217	0·8077
27·0	1·0177	0·9960	0·9751	0·9552	0·9361	0·9176	0·9000	0·8831	0·8667	0·8508	0·8356	0·8210	0·8068	0·7930
26·5	0·9989	0·9775	0·9571	0·9375	0·9183	0·9006	0·8834	0·8667	0·8506	0·8350	0·8202	0·8058	0·7918	0·7783
26·0	0·9800	0·9591	0·9390	0·9198	0·9014	0·8836	0·8667	0·8504	0·8346	0·8193	0·8047	0·7906	0·7769	0·7636
25·5	0·9612	0·9406	0·9209	0·9021	0·8841	0·8666	0·8-67	0·8340	0·8185	0·8035	0·7892	0·7753	0·7617	0·7489
25·0	0·9423	0·9222	0·9029	0·8844	0·8668	0·8497	0·8334	0·8177	0·8025	0·7878	0·7737	0·7601	0·7470	0·7343
24·5	0·9235	0·9037	0·8848	0·8667	0·8494	0·8327	0·8167	0·8013	0·7864	0·7720	0·7583	0·7449	0·7321	0·7196
24·0	0·9046	0·8853	0·8668	0·8490	0·8321	0·8157	0·8000	0·7849	0·7720	0·7562	0·7428	0·7297	0·7171	0·7049
23·5	0·8858	0·8668	0·8487	0·8313	0·8147	0·7987	0·7833	0·7686	0·7543	0·7405	0·7273	0·7145	0·7022	0·6902
23·0	0·8669	0·8484	0·8306	0·8136	0·7974	0·7817	0·7667	0·7522	0·7383	0·7247	0·7118	0·6993	0·6872	0·6755
22·5	0·8480	0·8299	0·8126	0·7959	0·7801	0·7647	0·7500	0·7359	0·7222	0·7090	0·6964	0·6841	0·6723	0·6608
22·0	0·8292	0·8115	0·7945	0·7783	0·7627	0·7477	0·7334	0·7195	0·7062	0·6932	0·6809	0·6689	0·6574	0·6461
21·5	0·8104	0·7930	0·7765	0·7606	0·7454	0·7307	0·7167	0·7032	0·6901	0·6775	0·6654	0·6537	0·6424	0·6315
21·0	0·7915	0·7746	0·7584	0·7429	0·7281	0·7137	0·7000	0·6868	0·6740	0·6617	0·6499	0·6385	0·6275	0·6168
20·5	0·7727	0·7561	0·7404	0·7252	0·7107	0·6967	0·6833	0·6704	0·6579	0·6459	0·6345	0·6233	0·6125	0·6021
20·0	0·7538	0·7377	0·7223	0·7075	0·6934	0·6797	0·6667	0·6541	0·6419	0·6302	0·6190	0·6081	0·5976	0·5874

Height of Barometer in inches (corrected).

(M.B.) I

TABLE IV.

Weight of 10,000 cubic feet of dry air in lbs. avoirdupois, at different Pressures and Temperatures.

Temperature.	0°	10°	20°	30°	40°	50°	60°	70°	80°	90°	100°	110°	120°	130°
31·0	896·3	877·2	858·8	841·3	824·4	808·2	792·7	777·7	763·2	749·3	735·9	723·0	710·5	698·5
30·5	881·8	863·0	845·0	827·7	811·1	795·2	779·9	765·1	750·9	737·2	724·0	711·3	699·0	687·2
30·0	867·4	848·9	831·1	814·1	797·8	782·1	767·1	752·6	738·4	725·2	712·2	699·7	687·6	675·9
29·5	852·9	834·7	817·3	800·6	784·5	769·1	754·3	740·0	726·3	713·1	700·3	688·0	676·1	664·6
29·0	838·4	820·6	803·4	787·0	771·2	756·1	741·5	727·5	714·0	701·0	688·4	676·3	664·7	653·4
28·5	824·0	806·4	789·6	773·4	757·9	743·0	728·7	714·9	701·7	688·9	676·6	664·7	653·2	642·1
28·0	809·5	792·3	775·7	759·9	744·6	730·0	715·9	702·4	689·4	676·8	664·7	653·0	641·7	630·8
27·5	795·1	778·1	761·9	746·3	731·3	717·0	703·2	689·9	677·0	664·7	653·8	641·4	630·3	619·6
27·0	780·6	764·0	748·0	732·7	718·0	703·9	690·4	677·3	664·7	652·6	641·0	629·7	618·8	608·3
26·5	766·2	749·8	734·2	719·1	704·7	690·9	677·6	664·8	652·4	640·6	629·1	618·0	607·4	597·1
26·0	751·7	735·7	720·3	705·6	691·4	677·7	664·8	652·2	640·1	628·5	617·2	606·4	595·9	585·8
25·5	737·2	721·5	706·5	692·0	678·2	664·8	652·0	639·7	627·8	616·4	605·4	594·7	584·5	574·5
25·0	722·8	707·4	692·6	678·5	664·8	651·8	639·2	627·2	615·5	604·3	593·5	583·1	573·0	563·3
24·5	708·3	693·2	678·8	664·9	651·6	638·8	626·4	614·6	603·2	592·2	581·6	571·4	561·5	552·0
24·0	693·9	679·1	664·9	651·3	638·3	625·7	613·7	602·1	590·9	580·1	569·7	559·7	550·1	540·7
23·5	679·4	664·9	651·1	637·7	625·0	612·7	600·9	589·5	578·6	568·0	557·9	548·1	538·6	529·5
23·0	664·9	650·8	637·2	624·2	611·7	599·6	588·1	577·0	566·3	556·0	546·0	536·4	527·2	518·2
22·5	650·5	636·6	623·4	610·6	598·4	586·6	575·3	564·4	554·0	543·9	534·1	524·8	515·7	506·9
22·0	636·0	622·5	609·5	597·0	585·1	573·5	562·5	551·9	541·7	511·8	522·3	513·1	504·2	495·7
21·5	621·6	608·3	595·6	583·5	571·8	560·5	549·7	539·3	529·3	519·7	510·4	501·4	492·8	484·4
21·0	607·2	594·2	581·8	569·9	558·5	547·5	536·9	526·8	517·0	507·6	498·5	489·8	481·3	473·1
20·5	592·7	580·1	567·9	556·3	545·2	534·5	524·2	514·8	504·7	495·5	486·7	478·1	469·8	461·9
20·0	578·2	565·9	554·1	542·8	531·9	521·4	511·4	501·7	492·4	483·4	474·8	466·4	458·4	450·6

Height of Barometer in inches (corrected).

APPENDIX B.

TABLE V.

Lifting power, in lbs. avoirdupois, of 10,000 cubic feet of Hydrogen (specific gravity 0·069), at different Pressures and Temperatures. It is assumed that the Temperature and Pressure of the Air are the same as those of Hydrogen.

Temperature.	0°	10°	20°	30°	40°	50°	60°	70°	80°	90°	100°	110°	120°	130°
31·0	834·3	816·5	799·5	783·1	767·4	752·3	737·8	723·9	710·4	697·5	685·0	673·0	661·4	650·2
30·5	820·8	803·3	786·6	770·4	755·0	740·2	725·9	712·3	699·0	686·3	674·0	662·2	650·7	639·4
30·0	807·3	790·1	773·7	757·8	742·7	728·0	714·0	700·6	687·5	675·0	662·9	651·3	640·0	629·2
29·5	794·0	777·0	760·8	745·2	730·3	715·9	702·1	688·9	676·1	663·8	651·9	640·5	629·4	618·7
29·0	780·5	763·8	747·9	732·5	717·9	703·8	690·2	677·2	664·6	652·6	640·8	629·6	618·7	608·2
28·5	767·1	750·5	735·0	719·9	705·5	691·6	678·3	665·6	653·2	641·3	629·8	618·8	608·0	597·7
28·0	753·7	737·3	722·1	707·3	693·2	679·5	666·5	653·9	641·7	630·1	618·8	607·9	597·4	587·2
27·5	740·1	724·0	709·2	694·7	680·8	667·4	654·6	642·2	630·1	618·8	607·7	597·1	586·7	576·7
27·0	726·6	711·1	696·3	682·0	668·4	655·2	642·7	630·6	618·8	607·6	596·7	586·2	576·0	566·2
26·5	713·2	697·9	683·4	669·4	656·0	643·1	630·8	618·9	607·4	596·6	585·6	575·3	565·4	555·8
26·0	699·7	684·8	670·5	656·8	643·7	631·0	618·9	607·2	595·9	585·1	574·6	564·5	554·7	545·3
25·5	686·3	671·6	657·6	644·1	631·3	618·9	607·0	595·6	584·5	573·9	563·5	553·6	544·0	534·8
25·0	672·8	658·4	644·7	631·5	618·9	606·7	595·1	583·9	573·0	562·6	552·5	542·8	533·4	524·3
24·5	659·4	645·3	631·8	618·9	606·5	594·6	583·2	572·2	561·6	551·4	541·5	531·9	522·7	513·8
24·0	645·9	632·1	618·9	606·2	594·2	582·5	571·3	560·6	550·1	540·1	530·4	521·1	512·0	503·3
23·5	632·5	618·9	606·0	593·6	581·8	570·3	559·2	548·9	538·7	528·9	519·4	510·2	501·3	492·8
23·0	619·0	605·7	593·1	581·0	569·4	558·2	547·5	537·2	527·2	517·6	508·3	499·4	490·7	482·2
22·5	605·5	592·6	580·2	568·4	557·0	546·1	535·6	525·6	515·8	506·4	497·3	488·5	480·0	471·7
22·0	592·1	579·4	567·3	555·7	544·7	533·9	523·7	513·9	504·3	495·1	486·2	477·7	469·3	461·4
21·5	578·6	566·2	554·4	543·1	532·3	521·8	511·8	502·2	492·8	483·9	475·2	466·8	458·7	450·9
21·0	565·1	553·0	541·5	530·5	519·9	509·7	499·9	490·6	481·4	472·7	464·1	456·0	448·0	440·4
20·5	551·7	539·9	528·6	517·8	507·5	497·6	487·0	478·9	469·9	461·4	453·1	445·1	437·3	429·9
20·0	538·2	526·7	515·7	505·2	495·2	485·4	475·1	467·2	458·5	450·2	442·1	434·3	426·7	419·4

Height of Barometer in inches (corrected).

Table VI.

Approximate Heights of the Barometer, corresponding to given Altitudes in feet. Also Volumes which 10,000 cubic feet of gas at 30 inches will occupy at other Pressures.

Col. 1. Altitude in feet.	Col. 2. Height of Barometer.	Col. 3. Volume occupied by gas.	Col. 1. Altitude in feet.	Col. 2. Height of Barometer.	Col. 3. Volume occupied by gas.	Col. 1. Altitude in feet.	Col. 2. Height of Barometer.	Col. 3. Volume occupied by gas.
		cub. ft.			cub. ft.			cub. ft.
0	30·00	10000	3700	26·19	11453	9800	20·93	14333
100	29·89	10036	3800	26·10	11495	10000	20·78	14437
200	29·78	10074	3900	26·00	11538	10200	20·63	14542
300	29·67	10113	4000	25·91	11580	10400	20·48	14649
400	29·56	10149	4100	25·81	11623	10600	20·33	14757
500	29·45	10187	4200	25·72	11666	10800	20·18	14866
600	29·34	10225	4300	25·62	11709	11000	20·03	14977
700	29·24	10260	4400	25·53	11752	11200	19·88	15091
800	29·13	10299	4500	25·43	11795	11400	19·73	15206
900	29·02	10338	4600	25·33	11838	11600	19·58	15322
1000	28·92	10374	4700	25·24	11881	11800	19·44	15432
1100	28·81	10413	4800	25·15	11924	12000	19·30	15544
1200	28·71	10449	4900	25·06	11967	12500	18·95	15831
1300	28·60	10490	5000	24·97	12010	13000	18·60	16129
1400	28·50	10525	5200	24·80	12097	13500	18·26	16429
1500	28·39	10566	5400	24·62	12185	14000	17·93	16732
1600	28·29	10604	5600	24·43	12275	14500	17·61	17036
1700	28·19	10642	5800	24·26	12366	15000	17·29	17351
1800	28·08	10682	6000	24·08	12458	15500	16·98	17668
1900	27·98	10722	6200	23·90	12552	16000	16·67	17996
2000	27·88	10761	6400	23·72	12648	16500	16·36	18337
2100	27·78	10800	6600	23·55	12739	17000	16·06	18680
2200	27·68	10839	6800	23·38	12832	17500	15·77	19023
2300	27·58	10878	7000	23·21	12925	18000	15·48	19380
2400	27·48	10917	7200	23·04	13019	18500	15·20	19737
2500	27·38	10957	7400	22·87	13117	19000	14·93	20094
2600	27·28	10997	7600	22·70	13215	19500	14·66	20464
2700	27·18	11038	7800	22·53	13313	20000	14·39	20848
2800	27·08	11079	8000	22·36	13413	21000	13·87	21629
2900	26·98	11120	8200	22·19	13515	22000	13·37	22438
3000	26·88	11161	8400	22·03	13618	23000	12·89	23274
3100	26·78	11202	8600	21·87	13717	24000	12·42	24155
3200	26·68	11244	8800	21·71	13818	25000	11·97	25063
3300	26·58	11286	9000	21·55	13921	26000	11·54	25996
3400	26·49	11328	9200	21·39	14025	27000	11·12	26978
3500	26·39	11369	9400	21·23	14128	28000	10·72	27985
3600	26·29	11411	9600	21·08	14230	29000	10·34	29013

APPENDIX B.

TABLE VII.

Calculation of Heights from Barometric Observations.

The difference in height between two stations is $(F - f)$ $\left(1 + \dfrac{T + t - 100}{1000}\right)$ where $F f$ are the heights corresponding to the barometer readings at the upper and lower stations respectively, and $T\, t$ are the temperature in degrees Fahrenheit.

Col. 1. Barometer reading.	Col. 2. Feet.	Col. 3. Difference for 0·01 in.	Col. 1. Barometer reading.	Col. 2. Feet.	Col. 3. Difference for 0·01 in.	Col. 1. Barometer reading.	Col. 2. Feet.	Col. 3. Difference for 0·1 in.
31·00	0	0	27·00	3765	10·1	23·00	8134	11·8
30·90	88	8·8	26·90	3866	10·1	22·90	8253	11·9
30·80	176	8·8	26·80	3968	10·2	22·80	8372	11·9
30·70	265	8·8	26·70	4070	10·2	22·70	8492	12·0
30·60	354	8·9	26·60	4172	10·2	22·60	8613	12·1
30·50	443	8·9	26·50	4275	10·3	22·50	8734	12·1
30·40	532	8·9	26·40	4378	10·3	22·40	8856	12·2
30·30	622	9·0	26·30	4481	10·3	22·30	8978	12·2
30·20	712	9·0	26·20	4585	10·4	22·20	9101	12·3
30·10	802	9·0	26·10	4689	10·4	22·10	9224	12·3
30·00	893	9·1	26·00	4794	10·5	22·00	9348	12·4
29·90	984	9·1	25·90	4899	10·5	21·90	9472	12·4
29·80	1075	9·1	25·80	5004	10·5	21·80	9597	12·5
29·70	1167	9·2	25·70	5110	10·6	21·70	9722	12·5
29·60	1259	9·2	25·60	5216	10·6	21·60	9848	12·6
29·50	1351	9·2	25·50	5323	10·7	21·50	9974	12·6
29·40	1444	9·3	25·40	5430	10·7	21·40	10101	12·7
29·30	1537	9·3	25·30	5538	10·8	21·30	10228	12·7
29·20	1630	9·3	25·20	5646	10·8	21·20	10356	12·8
29·10	1724	9·4	25·10	5754	10·8	21·10	10484	12·8
29·00	1818	9·4	25·00	5863	10·9	21·00	10613	12·9
28·90	1912	9·4	24·90	5972	10·9	20·90	10742	12·9
28·80	2007	9·5	24·80	6082	11·0	20·80	10872	13·0
28·70	2102	9·5	24·70	6192	11·0	20·70	11003	13·1
28·60	2197	9·5	24·60	6303	11·1	20·60	11135	13·2
28·50	2292	9·5	24·50	6414	11·1	20·50	11268	13·3
28·40	2388	9·6	24·40	6526	11·2	20·40	11402	13·4
28·30	2484	9·6	24·30	6638	11·2	20·30	11537	13·5
28·20	2580	9·6	24·20	6750	11·2	20·20	11672	13·
28·10	2677	9·7	24·10	6863	11·3	20·10	11808	13·6
28·00	2774	9·7	24·00	6976	11·3	20·00	11945	13·7
27·90	2871	9·7	23·90	7090	11·4	19·90	12082	13·7
27·80	2969	9·8	23·80	7204	11·4	19·80	12220	13·8
27·70	3067	9·8	23·70	7319	11·5	19·70	12358	13·8
27·60	3165	9·8	23·60	7434	11·5	19·60	12497	13·9
27·50	3264	9·9	23·50	7549	11·5	19·50	12637	14·0
27·40	3363	9·9	23·40	7665	11·6	19·40	12778	14·1
27·30	3463	10·0	23·30	7781	11·6	19·30	12919	14·1
27·20	3563	10·0	23·20	7898	11·7	19·20	13061	14·2
27·10	3664	10·1	23·10	8016	11·8	19·10	13204	14·3

(MB)

INDEX.

Subject.	Page.
A.	
Air, dry, table of weights of	114
Archimedes, principle of	43
Artillery fire at balloons	61
Astronomy, knowledge of, required	59
Atmosphere, data regarding	76
B.	
Ballast, output, formula of	45
Ballast to be retained for landing	54
Balloons, sizes of	24
Barometric observations for calculation of altitudes	117
Barometric wave, diurnal	76
Beaumont, Captain, R.E., report by	9
Bechuanaland, expedition to	14
Bernard, General, opinion of	14
Boiling point of water, heights from	77
Boyle's law	43
C.	
Capillarity, correction for	77
Captive work	33
Car, description of	26
Charles, Professor	6
Circle, properties of	90
Circles, table of	91
Coal gas	28
Clouds, varieties of	77
Collar, valve	24
Coutelle, Captain	6
D.	
Drill, balloon	34
Dip and distance of horizon	86
E.	
Envelope, construction of	21
Equilibrium, rupture of	45
Espitallier, Captain	7

INDEX.

Subject.	Page.

F.

Fleurus, battle of	7
Formula, barometric, to find difference in altitude	108
Formula for calculating dimensions and strength of tubes	17
Formula of ballast output	45
Formula of ballast to be retained for landing	54
Free runs	43
Free run book	56

G.

Galien, Joseph, pamphlet by	5
Gambetta	14
Gases, expansion of	43
Gases, specific gravity of	80
Gases, specific gravity of, by "Schilling's" apparatus	81
Gases, lifting power of	110
Generators, field	16
Gore, calculation of	21
Graphic determination of lifting power	109
Grapnel, description of	26

H.

Heights of barometer corresponding to given altitudes	116
Heights calculated from barometric observations	117
Heights from boiling point of water	77
Hoop, neck	24
Hoop	25
Horizon, dip and distance of	86
Hot-air balloons	28
Hydrogen first used in ballooning	6
Hydrogen, preparation of	28
Hydrogen, compressibility of	43
Hydrogen, table of lifting power	115

I.

Instructions for conduct of free run	50
Imperial and metric systems, comparison of	88

L.

Lana, Francis, project of	5
Laws of practical ballooning	53
Lifting power, graphic determination of	109
Lightning, precautions against	42
Logarithms of numbers	102
Liquids, specific gravity of	80

Subject.	Page.
M.	
Mariner's compass, points of	87
Metals, specific gravity of	80
Metric system	88
Miscellaneous tables	75
Montgolfier Brothers, invention of hot-air balloons by	5
Moral effect of balloons in war	62
N.	
Napoleon in Egypt	8
Net, construction of	25
P.	
Paris, Siege of	14, 61
Photography from balloons	71
Points of mariner's compass	87
Practical ballooning, laws of	53
R.	
Reconnaissance, free run	63
Richmond, balloon at	11, 60
Ring, net	25
Rope, rules for weight and strength of	84
Rope, wire	26, 84
Rope, grapnel	26
Rupture of equilibrium	45
S.	
Schilling's apparatus	81
Signals from a balloon	37, 60
Sketching from a free balloon	65
Slack balloon, law regarding	48
Specific heat of air and H. compared	49
Specific heat of various substances	82
Spheres, table of	97
Sphere, properties of	96
Suakim, balloons at	15
Sulphuric acid and zinc process	31
T.	
Tables, explanation of Lieut.-Colonel Watson's	104
Tables, miscellaneous	75

INDEX.

Subject.	Page.
Telephone communication from car	41
Telegraph wires, crossing	39
Thermometers, comparison of	83
Trailing rope, use of	54
Trees, passing	38
Tube, gas, used in British service	17
Tubes, rules for loading and testing	17
Tubes, formula for calculating dimensions of	17

V.

Valve tube	19
Valve top	25, 33

W.

Wagons, balloon section	27
Wagons, weight of	27
War, employment of balloons in	59
Watson's tables	104
Weights and specific gravities	80
Wind, rate of, in captive work	33
Wind pressure, table of	75

Z.

Zinc and sulphuric acid process	31